WATER

SCIENCE <u>1995</u>

Matt — so much to know! — I
wish I could be there to enjoy
this book with you. Living in
the desert here, I really think about
water - it is <u>so</u> important! Hope
you enjoy this — the <u>weather</u>
brings us <u>water</u>. PEG.

Many good Experiments in here.

WATER SCIENCE

WRITTEN BY **DEBORAH SEED**
ILLUSTRATED BY **BOB BEESON**

Addison-Wesley Publishing Company, Inc.

Reading, Massachusetts ○ Menlo Park, California ○ New York
Don Mills, Ontario ○ Wokingham, England ○ Amsterdam ○ Bonn
Sydney ○ Singapore ○ Tokyo ○ Madrid ○ San Juan
Paris ○ Seoul ○ Milan ○ Mexico City ○ Taipei

Many of the designations used by manufacturers and sellers to distinguish their products are claimed as trademarks. Where those designations appear in this book and Addison-Wesley was aware of a trademark claim, the designations have been printed in initial capital letters (e.g., Joy, Nintendo).

Neither the Publisher nor the Author shall be liable for any damage which may be caused or sustained as a result of conducting any of the activities in this book without specifically following instructions, conducting the activities without proper supervision, or ignoring the cautions contained in the book.

Library of Congress Cataloging-in-Publication Data

Seed, Deborah.
   Water science / written by Deborah Seed ; illustrated by Bob Beeson.
      p.   cm.
Includes index.
Summary: Discusses the functions, properties, and vital importance of water in our lives and examines what water hazards we face as we misuse or pollute it. Includes water games and tricks.
ISBN 0-201-57778-X (alk. paper)

1. Water – Juvenile literature. [1. Water.] I. Beeson, Bob, ill. II. Title.

QC920.S38 1992   553.7 – dc20   92-9833
                                                    CIP
                                                    AC

Edited by Valerie Wyatt
Interior design by Christine Toller
Cover by Fred Harsh
Set in 12-point Century Expanded by Leading Type

Originally published as *The Amazing Water Book* by Kids Can Press, Ltd., of Toronto, Ontario, Canada.

1 2 3 4 5 6 7 8 9-AL-95949392
First printing, May 1992

Addison-Wesley books are available at special discounts for bulk purchases by schools and other organizations. For more information, please contact:

Special Markets Department
Addison-Wesley Publishing Company
Reading, MA 01867
(617) 944-3700 x 2431

Text stock contains over 50% recycled paper

For Peggy, the gardener, and Jack Seed,
the pond-maker, and Thomas and Catherine,
water enthusiasts.

Acknowledgements

The author is grateful to the following people for helping with the research and checking the accuracy of the manuscript: Carl Andersen, Department of Oceanography, Dalhousie University; Tony Seed, New Media Services, Halifax; Bob Semple, professional diver and underwater photographer, Halifax; Pam Geddes, underwater hockey player, Halifax; Michelle Roy, Bedford Institute of Oceanography, Dartmouth, N.S.; Peel Board of Education science teachers Frank Gold, John Perkins, Punita Kandasamy and geography teacher Bruce Savage, Mississauga; Dr. Christa Jeney, Toronto pediatrician, Dr. Marjon Blouw and Dr. Rodney Drabkin, Victoria physicians, Dr. Robert Williams, veterinarian, and Kathy Williams; Margaret Griffin, co-author of *The Amazing Egg Book* and researcher *par excellence*, and Ruth Griffin, retired science teacher; canoeist Linda Collier, and researchers/copy checkers Elaine Freedman, Edna Barker and Martin Reyto. These institutions also provided invaluable resources: Westwood Secondary School Library, Mississauga; Runnymede Library, Toronto, and Richview Library, Etobicoke; Scripp's Institute of Oceanography, California; Ontario Ministry of the Environment, Pollution Probe, and Ontario Hydro. I am likewise indebted to Thomas, Catherine and their friends for testing the activities and tricks and rejecting those deemed boring to kids aged nine to twelve. A big thanks also to Wendy Thomas, copyeditor, Val Wyatt, editor, and the crew at Kids Can Press for their assistance.

WATER

SCIENCE

Contents

*Water, water, what a pain!
Always going down the drain.*

1. You are all WET!

Pinch yourself. You feel pretty solid, don't you,
but you're not. About two-thirds of your weight
is water. If you're tall and thin, your body
might contain as much as 70% water; if you're
chubby, your body might have much less.
You're not a human being—you're a water being!

In a sense, you came into the world as a water baby. For nine months, as you grew within your mother's womb, you floated around in a small swimming pool called an amniotic sac. This sac protected you from infections and from being jostled around too much. When you were born, the amniotic sac broke open and released its water.

The amount of water in your body changes as you grow. A newborn baby has the highest water content — about 78%. Your water level falls as you get older and your bones and tissues develop. By the time you become an adult, you'll only be about 60% water.

You can see the water in your body when you bleed, urinate (pee) or sweat. Even

We are mostly water!

your hardest parts, such as your bones and teeth, contain some water.

Water makes up 80% of your blood. Blood carries food and oxygen to the millions of tiny cells throughout your body. The cells use the food and oxygen to give you energy and to help you grow. Waste is produced in the process. Your blood carries this waste away from the cells. The waste leaves the body in urine or sweat or even tears, all of which are mostly water.

Water also helps to control your body temperature so that you don't get overheated. And water lubricates your muscles so they can move. Without water, you couldn't even bat an eyelash.

Jobs water does in the body

- Tears keep the eyes clean and allow you to blink.
- Lymph fluid, mostly water, flushes out bacteria and germs.
- Saliva moistens food and contains chemicals that help break it down before it goes to the stomach. Stomach juices help to digest the food.
- Blood carries digested food and oxygen to the cells and removes wastes and carbon dioxide.
- Water-filled kidneys filter impurities from the blood. The wastes come out in urine.
- Sweat gets rid of excess water and allows the body to cool itself.

Have you ever heard of anybody going on a water strike? People go on hunger strikes to draw attention to their causes, but no one dares to go without water. People can live without eating for a few weeks, as long as they keep drinking. But if they run out of water, they would probably die in less than a week. Why couldn't they last any longer? Their bodies would dry out.

When your body's water supply gets too low, the thirst centre in your brain sends out a message to the back of your throat: get a drink FAST! Your sponge-like body is getting dry! Every day you need about 2 L (3 pints) of water to make up for the water you lose in sweating, breathing and peeing. Drinks, such as milk, supply you with nearly half the water you need. The rest of the supply comes from the food you eat.

All food contains some water. The driest foods are seeds, which are between 5% and 10% water. The wettest foods are vegetables and fruit. The best thirst-quencher of all is watermelon, because it's mostly water.

Drink up your lunch

You get rid of the same amount of water that you drink or eat every day. How? A little water is expelled in your solid waste, which doctors call a bowel movement. More than half the water you consume comes out in urine, and the rest is sweated or breathed out. Want to see some of the water you expel every time you breathe? Blow on a mirror for a few seconds. What happens? Water vapour in your warm breath changes to liquid water when it hits the cold surface of the mirror.

milk: 87% water

sunflower seeds: 5% water

carrot and cucumber slices: 90% water

apple: 80% water

sliced-up tomato: 94% water

sandwich: 35% water

7

Cool it!

Sweating helps your body cool down. When it's hot outside or when you have a fever, your blood vessels widen so that the warm blood can travel to the two million sweat glands located on the surface of your skin. The warm blood makes the sweat glands release water. You look flushed but feel cooler. The sweat evaporates into the air, taking the excess heat away from your body. Phew! That's better!

When sweating doesn't cool you off enough in the summer, help your body get rid of more heat by making the towels sweat for you. Dampen two large towels, lie down in a shady spot and cover yourself with the wet towels. Water will evaporate from the towels and help take the heat away from your body. (Putting a moist towel over your forehead is a good way to cool off a fever, too! Watch that you don't shiver, though. Shivering *raises* your internal temperature and decreases sweating.)

You lose about 3 glasses (less than half a litre or 1 pint) of water every day through perspiration. You lose far more water on hot days or when you're sick with a fever. You need to drink lots of fluids to replace this lost water! Try slurping down a slush. Put a can of orange- or grape-juice concentrate, two cans of water and some ice cubes in the blender. Whiz the mixture for a few seconds. Add a can or glass of ginger ale to give your slush some sparkle. Start sipping.

The rivers within you

A complicated system of rivers in your body circulates the blood. These small rivers are called arteries and veins, or blood vessels. The heart pumps fresh blood through the widest rivers, the arteries, to take blood up to the brain and down to the arms and legs. These arteries form smaller and smaller creeks called capillaries, which are tiny tubes that take blood to all the tissues and the skin. Another system of rivers, called veins (they're the blue lines on your wrists), takes the blood back to the heart to get a fresh load of oxygen from the lungs.

What a wonderful system!

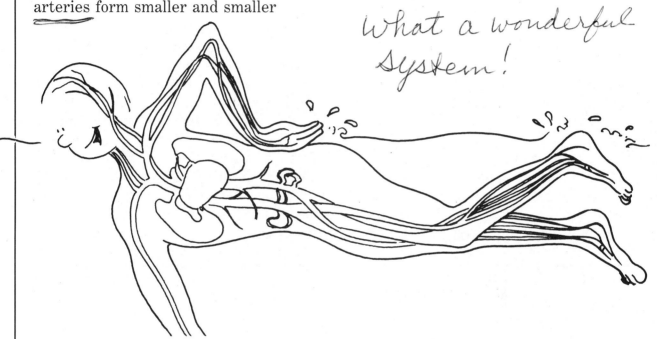

The ocean within you

Why do your tears and blood taste salty? Because you've got a little bit of the ocean inside you. The ocean was the home of the first plants and animals. When the creatures came ashore to live on land millions of years ago, they carried some of the sea within their bodies. That's why humans and other creatures have salty water in their bodies.

Surviving on little water

Like people, all animals and plants need to maintain the water level in their bodies to survive. Fish and aquatic plants have it easy — they live in water all the time. But land plants and animals are not so lucky. They must find water wherever they can. Desert creatures have it really tough. Because it rains so seldom in deserts, the plants and animals living there must go for weeks, months, even years without water. How to they do it?

To avoid overheating, many desert animals hide under boulders or in burrows during the day, hunting for food at night or in the early morning. Deep underground burrows are cool, and the animals' breath also keeps them moist. The champion burrow-maker is the spadefoot toad of North America: with its spade-like hind legs, it digs more than a storey down, where it hibernates until rain finally falls.

Desert animals can't afford to sweat either, because their bodies would lose water that's hard to replace. How do some get rid of their excess heat? Through their ears. The tiny fennec fox and the desert rabbit have enormous ears that act like radiators to eliminate body heat carried there by blood vessels.

Many desert animals have developed ways to store water. Where does the camel, the champion waterstorer, conserve its water? Not in its hump, where it keeps its food reserves in the form of fat. Its water is stored in its stomach. A camel can drink up to 80 L (70 quarts) of water at one time, one-third of its body weight. To avoid losing any water, the camel even swallows its own nose drippings and discharges feces so dry that people collect them to build fires. If its water storage runs low, the camel can turn part of the fat in its hump into liquid.

Desert plants have also come up with clever ways to conserve water. Cacti have waterproof stems that act like water-storage tanks. The giant saguaro cactus can absorb a tonne (ton) of water in one day and store it in its stem until the next rainfall. Like many desert plants, it doesn't have leaves from which water can escape. Desert plants that *do* have leaves have waxy coatings to stop water from escaping through evaporation.

Many desert plants spend their lives as dormant seeds lying in wait for a rainfall. If it rains, the seeds sprout at once and produce leaves, flowers and seeds very rapidly. For a few short days, the desert blazes with brilliant colour. Then the plants die and the seeds lie in wait for the next rainfall.

amazing cactus!

Thirsty plants

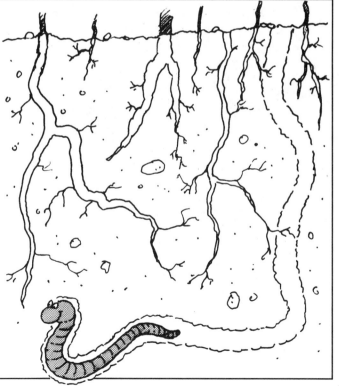

Land plants send down roots to draw water from the soil so they can make food. The bigger and taller the plant, the bigger the roots. Can you guess how much water a tall tree needs to stay healthy? Answer on page 103.

Plants need water for another important reason: it holds them upright. When you fill a water balloon with water, it becomes very stiff. A plant contains thousands of balloon-like cells. Filled with water, these cells keep the plant stiff. What happens when the cells lack water? The leaves droop, the stem wilts and soon the whole plant flops over. After a rainfall, the cells fill with water again. Up goes the plant.

Desert champs

The kangaroo rat of North America doesn't need to drink *or* store water. Sound impossible? It lives on hard dry seeds and *makes* its own water as it digests its food.

Some animals lick dew off plants or eat leaves to get water. Desert beetles stand on their heads so that the water droplets from fog trickle down into their mouths.

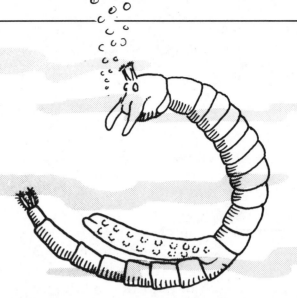

Fairy shrimp eggs can survive for 100 years without water and still hatch after a rainfall. They mature and lay new eggs before the pools dry up again.

All birds need water — even desert birds. The sand grouse of Africa finds water wherever it can and transports it to its young in its nest. How? The male has special sponge-like feathers on its belly that it soaks in water. Back at the nest, the chicks drink the water from the male's feathers.

Pets need water, too

On hot days, dogs don't perspire to get rid of their body heat — they pant. The heat escapes through blood vessels in the tongue. A panting dog needs water.

To check if your cat needs water, gently pinch the scruff of its neck (the scruff is the skin between the shoulders and neck, where you pick up a kitten). Twist your hand to the right and let go. The skin should go right back down again. If it takes more than one second to go down, the cat needs to see a veterinarian.

Keep your pet supplied with a bowl of fresh water every day, so it won't need to drink out of toilets or puddles!

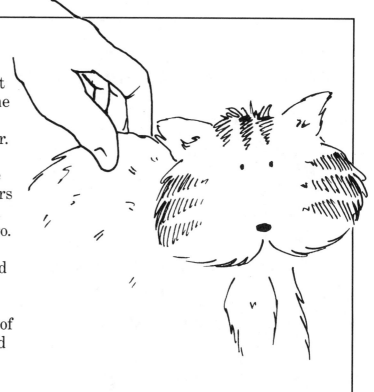

Did you know?

A jellyfish is 95% water, a dog is 70% water and a frog is 78% water.

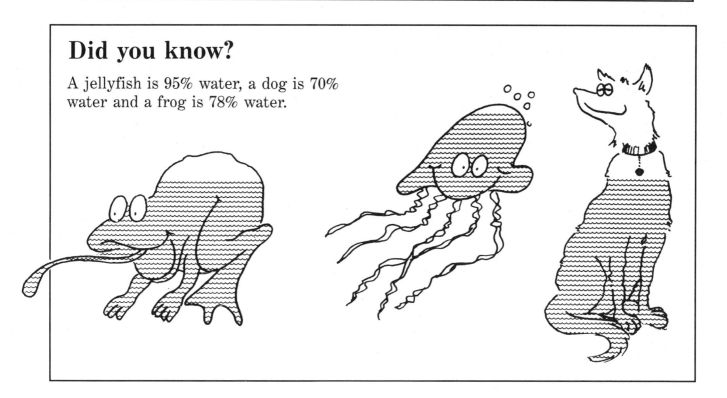

2. Wonders of water

Suppose you turned on a tap and out gushed orange juice instead of water. Imagine brushing your teeth or washing your hair with orange juice! Could you wash the dishes or water the lawn?

We're so used to water that we forget how unusual it really is. Water is the only substance that:
- we can drink, bathe in and water plants with
- can be found on Earth as a solid (ice), liquid or gas (water vapour)

- is colourless, odourless and tasteless in its purest state.

Water may seem ordinary, but it has some amazing abilities that would make Superman and Wonder Woman jealous. Discover some of the wonders of water by trying out the experiments in this section.

H₂O = water

If you could shrink small enough to step inside a drop of water, you'd find yourself surrounded by a crowd of teddy-bear heads. Each of these tiny heads is a water molecule, and they're so tiny you couldn't see them under a powerful microcope. Molecules are made up of even smaller bits called atoms. In a water molecule, the face of the teddy-bear is an oxygen atom, and the two ears are hydrogen atoms.

Be prepared for a rough ride in your water drop if the sun comes out. Heat energizes the water molecules, making them go wild. The molecules race around like out-of-control speeding trains. When they collide, the impact is so strong that many molecules are shot right out of the water drop into the air. They become water vapour.

If the temperature cools down, beware. Cold slows down the movement of the water molecules and packs them more tightly together. If it gets *freezing* cold, the water molecules will start bonding together. Soon you'll be surrounded by strong six-sided ice crystals. Step out of the drop before you're frozen, too!

Hydrogen and oxygen are invisible gases. When they combine in just the right way — two parts hydrogen for every one part of oxygen — they produce water. This is why scientists have given water the chemical symbol H_2O: H_2O means that each water molecule contains two H's (hydrogen atoms) and one O (an oxygen atom).

The discoverer of water

For hundreds of years scholars thought that water was an elementary substance that couldn't be broken apart. Just 200 years ago, a British scientist called Henry Cavendish discovered that water was really a compound of two common gases. Imagine the shocked reaction of the scientists when Cavendish heated up the right mix of hydrogen and oxygen and produced water!

Water, the transformer

Is it a solid? Is it a liquid? Is it a gas? It's all three. Water is nature's greatest transformer. It can change from one form to another and back again. How? Try these experiments and see.

Quick change

You'll need:
a pot filled with water
a stove
some oven mitts
a cold spoon
an empty glass

1. Heat the water on the stove. Get an adult to help you. See the bubbles rising to the surface? These are bubbles of water vapour. Don't put your face too close to the pot — when these bubbles escape, they're hot and can burn you.
2. Put on the oven mitts and hold the cold spoon over the water vapour rising from the pot. The spoon's cold surface will transform the water vapour back into liquid water. Catch the water droplets in the glass.

You transformed liquid water into a gas (water vapour) and back into liquid water. How did you do it? By using heat and cooling. The molecules in liquid water are constantly moving around, bumping into each other like bumper cars at a fair. When you heat the water, you make the molecules move faster and faster, like speeding trains. Finally, some of them shoot right out of the pot and into the air. They become water vapour, an invisible gas.

This process of changing from a liquid to a gas is called evaporation. (See the word "vapour" buried in the word "evaporation"?) Evaporation occurs at the surface of water when the molecules escape into the air as water vapour. Boiling is a special case of evaporation: water vapour doesn't just escape from the surface. It escapes from the bottom, middle and top. That's why you see the lines of bubbles in your pot.

When you cooled the water, you reversed the process. Cooling water vapour into liquid water is called condensation.

Water is constantly changing from one form to another everywhere on Earth. The heat from the sun makes water evaporate from the surface of lakes and oceans. The water vapour rises into the air, cools and condenses to form clouds. Water's transformations give us rain and snow. See pages 46-48 for more about water and weather.

Feeling foggy?

Breathe out on a cold winter day and you'll make a fog. Fog is formed whenever warm, wet air (such as your breath) hits colder air (such as the outside air on a chilly day). Presto — out come trillions of tiny drops of water. Fog is simply a low cloud, a wet blanket you can see.

Pop goes the stopper!

What happens when you cool water even more? Try this and see.

You'll need:
a plastic milk jug
some aluminum foil
a pan large enough to hold the jug

1. Fill the jug to the top with water.
2. Wrap a piece of foil over the top to act as a stopper.
3. Put the jug in the pan, and place the pan in the freezer overnight. What happens to the foil stopper after the water freezes?

Freezing liquid water transforms it into ice. But here's where water is unique. Most liquids contract, or shrink, when they cool and turn into a solid. Hot liquid wax from a candle, for example, becomes a smaller blob of solid wax when it cools. Water does the opposite when it turns into a frozen solid — it expands and takes up *more* space. If anything is in the way, such as the foil stopper, a water pipe or a sidewalk, ice pushes against it to get the room it needs. And ice is strong — it's powerful enough to split rocks apart.

(N)ice trick

Ice has another trick that makes it unique. It floats. Why? When water molecules freeze, six molecules lock together to form an ice crystal. Inside the crystal is empty space. Ice floats because it's less dense than water.

It's a good thing that ice is lighter than water, because it means that lakes and ponds freeze only at the top. The ice forms a protective blanket over the pond — the water underneath doesn't freeze, and fish and other water animals can survive through the winter. If water did behave like other substances, the solid ice would sink to the *bottom* of ponds rather than float to the top. Ponds would freeze from the bottom up, one layer after another, killing all the creatures in the process. And the bottom layers would remain as ice throughout the summer!

18

Cool sports

If you think skating, skiing and curling are snow and ice sports, guess again. They're all water sports. When you skate, you're really travelling on a thin film of water. The weight of your body presses down on the ice and causes the ice underneath your blades to melt, making the surface extremely slippery. On bitterly cold days, you'll have a harder time skating outdoors because the pressure of the skates can't melt the ice.

Yes, skiing on snow is a water sport, too. The snow is melted by the friction (rubbing) of your skis. On cold days, there isn't enough friction to melt the snow underneath wooden skis, so that's why you have to wax them to make them slide better.

Ever wonder why curlers sweep the ice so hard in front of the rocks thrown by their team-mates? They're making the ice melt to form a thin film of water, so the rocks will slide farther.

19

Water, the dissolver

Water can make things disappear. Don't believe it? Put a spoonful of sugar into a glass of water and stir. Before long, the sugar has vanished before your eyes. You can't see the sugar molecules because they've separated and spread evenly throughout the water. But you know the sugar is still in the water because the water tastes sweet. What's happened? The water molecules have ripped apart the molecules that are holding the sugar together. The sugar molecules are now surrounded by water — and so are dissolved.

Water has the special ability to make things disappear, or "dissolve," which is why it is such a useful solvent. Over time, water can dissolve minerals in the soil for plants to use; gases, such as oxygen, for fish to breathe; and even air-borne chemicals that bring us acid rain. Water can't dissolve all substances, such as oils and fats and waxes, but it's one of the best solvents around.

The disappearing act

Try this activity to see which substances really dissolve in water and which are only suspended in it. If the substance has dissolved, the solution (the mixture of water and the dissolved stuff) should look the same throughout. The particles will have broken up and will be spread evenly in the water. If the mixture stays cloudy, and the particles hang there and then settle to the bottom, you've made a "suspension" rather than a "solution."

You'll need:
5 drinking glasses
5 mL (1 teaspoon) each of salt, flour, baking soda, sugar, dirt

1. Fill the glasses with cold water.

2. Stir a spoonful of salt into the first glass.
3. Stir a spoonful of flour into one glass, baking soda into another, and so on.
4. Wait for a few minutes to see what happens.

Which solids disappear, or dissolve, in the water? Which are suspended in the water and eventually settle to the bottom? Like the baking soda in the glass, bits of rock and soil can be suspended in fast-moving rivers and streams. When the water flows more slowly, the dirt and rock settle to the bottom of the stream bed.

How much sugar can you dissolve in a glass of cold water? Keep adding one small spoonful after another. When no more sugar can be dissolved, the solution has become saturated. That's the way you feel after eating too much fudge and candy. No more, please!

Try this experiment again with a glass of very hot water. Can you dissolve more sugar in hot water than in cold water? How much more? Heat allows water to dissolve a larger amount of the solids. That's why you heat milk and water to make soups and sauces.

ock candy

This experiment is so simple, it's sweet.
WARNING: Avoid a steam burn — don't put your face too close to the pan. Wear oven mitts.

You'll need:
250 mL (1 cup) water
a pan
375 to 500 mL (1½ to 2 cups) sugar
a glass jar
a piece of thick string, wool
   or pipe cleaner
a spoon to hang the string from

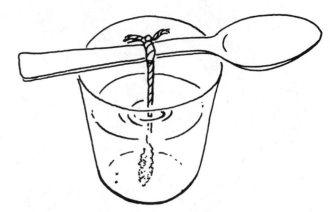

1. Heat the water in a pan until it boils. Stir in the sugar gradually until all of it has dissolved. The mixture will become a thick syrup.
2. Pour the syrup into a glass jar.

3. Tie one end of the string to the middle of a spoon. Rest the spoon on the jar rim so that the string hangs into the liquid.
4. Leave the jar on a counter for a few days. What happens?
   As the mixture cools and the liquid evaporates, some of the dissolved sugar comes back out of the water in the form of crystals. Enjoy your snack!

Salt garden

Crystals form in nature all the time. In snowy areas, winter boots get marked with crust-like white patches. These are salt crystals made by road salt left behind after the water's evaporated. In hot areas beside the ocean, people rope off large shallow beds to collect salt crystals from sea water. The salt is left behind once the water evaporates in the hot sun.

Want to make an unusual rock garden? Put your favourite stones in a shallow dish. Make a super-salty solution by dissolving as many spoonfuls of salt as

you can into a glass of warm water. Pour your salt solution over the rocks and put the dish in a sunny window. After a few days, you'll be amazed by how the salt crystals transform your favourite rocks.

20, 21, 22

Drink some air

Water dissolves gases, too. If you place a glass of cold water in a warm place for a day or so, you'll see bubbles of air rising to the surface. These air bubbles contain oxygen that was dissolved in the cold water. As the water warms up, the oxygen comes out. That's why many types of fish and aquatic animals thrive in cold water that's rich in dissolved oxygen and die in warm water.

Caves and salt icicles

Water can even dissolve some types of rock, such as limestone, to make spectacular underground caves. How? Falling rain first dissolves carbon dioxide in the air, which makes the rain mildly acidic. This weak acidic solution then dissolves limestone as it moves down through the cracks in the rocks. The result: caves.

Some limestone caves have become huge tourist attractions because of the rock columns that hang down from the ceiling (stalactites) or grow up from the cave floor (stalagmites). These strange rock icicles are formed by the minerals left behind by the evaporated water.

The tallest stalagmite in the world is about 30 m (98 feet) tall, that's 10 storeys tall. This salt giant is in the Aven Armand cave in Lozère, France. Scientists say that these columns grow about ¼ mm (1/100 inch) a year, so how long has this column been growing? Answer on page 103.

23

Water, the attractor

Wet two pieces of plastic bag and hold them together. Now try to pull them apart. The plastic pieces want to stick together. They're being "glued" to one another by the water. Water molecules are strongly attracted to each other and hold on for dear life. Put them together and they want to stay together. The strong attraction water molecules have for one another lets water do some surprising tricks.

Shoot some starch

Here's a dramatic way to break water's tough skin.

You'll need:
a shallow dish
some cornstarch or talcum powder
a toothpick or knife
detergent

1. Fill the dish with more than 2.5 cm (1 inch) of water.
2. Sprinkle some cornstarch or talcum powder evenly over the surface.
3. Dip the toothpick or knife in some detergent. Touch the centre of the dish with this soapy tip. What happens?

Water's tough skin will hold up cornstarch or talcum powder. But adding the detergent instantly weakens the skin, breaking the tension. The cornstarch shoots to the other side of the dish.

Water's t-e-n-s-e skin

How tightly do water molecules cling to one another? Try overfilling a full glass of water to find out.

You'll need:
an eye dropper or turkey baster
a bowl of water
a glass of water filled to the top

1. Fill the eye dropper or baster with water from the bowl by squeezing the rubber end.
2. Now carefully release one drop of water at a time into the full glass of water.
3. Keep adding as many drops as you can until the surface bulges over the top of the glass.

As you add more and more drops, the surface of the water stretches. The attraction of water molecules is so strong at the surface that they hold onto each other even when the glass is overfilled. The water looks as if it has a skin over it. This is called surface tension. Eventually, you'll add too many drops — the water will spill because the tension has broken.

Many insects make use of water's surface tension to walk, skate and ski across water. The next time you go to a pond, take along an insect book and see how many skaters you can identify. Look for these insects.

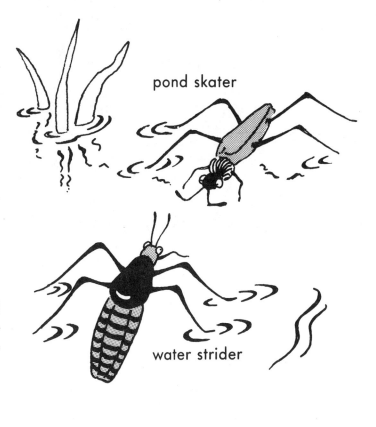

pond skater

water strider

Blowing bubbles

When you blow bubbles, you use soap to stretch water's tough skin and make it super-elastic. The best place to experiment with monster bubbles is outside on a *humid* day. That way you'll be able to enjoy seeing the rainbow in the soap film and not have to worry about making a big mess. Remember: dryness is a bubble's enemy. When a large bubble hits a dry surface, it'll explode. Be careful that soap doesn't get into your eyes.

Best bubble brew

Mix up a batch of this brew for some terrific bubbles.

You'll need:
125 mL (½ cup) thick dishwashing
   detergent such as Joy
1 L (1 quart) water
string
straws

1. Pour the detergent and water into a pan and mix.

2. To make a bubble frame, thread some string through two straws and knot.

3. Dip the square frame into the bubble brew. Pull the frame out slowly so the soap film doesn't break.

Try these tricks with your frame:

- Lift the frame up to your face and blow the soap film gently to make a large long bubble.

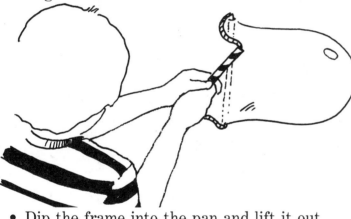

- Hold the frame vertically and slowly pull it through the air. Can you make an even longer bubble?

- Dip the frame into the pan and lift it out holding the straws together. Gently pull the straws apart, lift up the frame and then bring the straws together.

- Bend a piece of wire into unusual shapes to make weird bubbles.

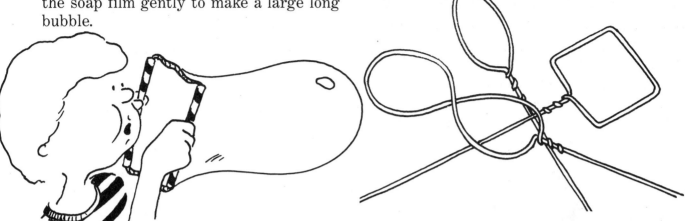

- Tape toilet-paper rolls or cardboard rolls together and blow through them to make monster bubbles.

Water, the mover

Gravity is the powerful, invisible force that pulls everything down toward the centre of the earth. That's why rain falls down to the ground, then seeps through the earth and rocks to collect as groundwater.

So how do plants and trees get water to climb from their roots right up to their leaves? In these activities, you'll discover other powerful forces at work in nature.

One of these "forces" is called capillary action. It's so powerful that a tree 60 m (200 feet) high can soak up four bathtubs of water in a single day.

Capillary action!

- Capillary action dries you after a bath. Water from your skin travels up between the fibres in the towel. Thick towels with long strands dry better than thin towels.
- Capillary action makes breakfast cereals soggy with milk.

- When you sit on damp ground, you'll feel the damp faster if you're wearing cotton pants or shorts than if you're wearing clothes made from polyester. Water shoots up between the open fibres of cotton faster than it goes up between the more solid fibres of polyester. Your best bet is to sit on a piece of plastic. It doesn't absorb any water at all.

- Surprise your friends by feeding them red-striped celery. The night before they come over, place a stalk of celery in a glass of water, and add a few drops of red food colouring. The next day when you cut the celery into pieces, notice the tiny red tubes filled with water.

- Want to give your parents a weird bouquet of flowers? Carnations are cheap to buy and work well. With a very sharp knife, cut each stem in two, making a straight vertical cut. Be careful here: if you slice cross-wise you might damage the tubes inside. Put each half stem in a different glass of water. Stir red food colouring into one glass and green into the other. A few hours later the petals will be two-toned.

Water bookmarks

See the magic of capillary action by making these bookmarks.

You'll need:
scissors
a piece of blotting paper or several
  coffee filters
water-soluble felt-tip markers
sticky tape
several small bowls filled with water

1. Cut the paper into long, thin strips for your bookmarks.
2. Use a felt-tip marker to put a big dot or design about 3 cm (1 inch) from the bottom of a strip.

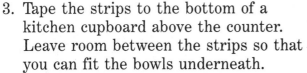

3. Tape the strips to the bottom of a kitchen cupboard above the counter. Leave room between the strips so that you can fit the bowls underneath.
4. Put some pans or books on the countertop to make a stand for the bowls, then place a bowl under each strip so that the end of the paper touches the water. Watch the water climb up the strips.

5. Let the strips dry out before putting them in your books.

Why does the water climb up the paper? Paper is made of tiny fibres that are pressed close together. (You can see them under a microscope.) Between the fibres are tiny tubes. Water molecules travel up these tubes. First the water molecules stick to the fibres so the water begins to rise, and then the other water molecules follow like links on a chain. This tendency of water to pull itself up tiny tubes is called capillary action. ("Capillaries" means tiny tubes.)

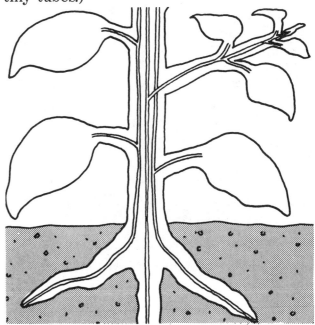

A chain of water molecules is yanked up through the narrow tubes in the roots and stems of plants in the same way. The plant needs the water and its dissolved nutrients to make its food. What happens when the molecules get to the plant's leaves? They evaporate out of tiny holes in the leaves. More molecules are pulled up to take their place, and the chain reaction continues. The plant moves water and its dissolved nutrients from the soil to the topmost leaves, where it makes its food.

Water, the "presser"

Put on a rubber glove and plunge your hand into a sink full of water. Does the water stick the glove to your hand? Water pressing on your gloved hand feels strange but not uncomfortable. But if you were to put on SCUBA gear and dive deep into an ocean or lake, the water pressure would soon feel uncomfortable. The pressure increases the deeper you go. In the deep ocean, the water pressure might be a thousand times greater than that at sea level. The pressure in these deep areas would be enough to crush you — or a submarine.

Deep-sea divers have to wear special diving suits and carry a pressurized air supply to protect themselves against the pressure of the water. Undersea vehicles (called submersibles and bathyscaphes) allow people to go much deeper than SCUBA divers. These vehicles are specially constructed to withstand the terrific pressure of water at very great depths. The bathyscaphe *Trieste* went to the deepest part of the Pacific Ocean in 1960, allowing the scientists aboard to explore life at 10 900 m (35 800 feet) below sea level.

How to make water squirt

The deeper the water, the stronger the pressure. Don't believe it? Try this.

You'll need:
a hammer and a nail
a large empty juice can
some tape

1. Use the hammer and nail to make three holes in a straight vertical line at the top, middle and bottom of the can. Put tape over the holes.
2. Place the can on the counter at the edge of the sink, with the taped holes facing the sink.
3. Fill the can with water and remove the tape.
4. Look at the water streams coming out of the holes. Which hole makes the longest stream of water? Why?

The bottom stream shoots out the farthest because it's being pushed down by the weight of the water above. The air pushes it down, too. This stream is under the highest pressure. The stream at the top is under the lowest pressure.

Hot and cold layers

When you stand beside the sea, the atmosphere pushes down on your body with a force of 1 kg each square cm (14.7 pounds per square inch). Air is heavy even if you can't see it pressing down on you. The combined weight of air and water makes it even harder to swim underwater.

When you stand in a pond or lake, you may notice that the water at the top of your body is warmer than the water around your feet. Do an underwater dive and the water is even colder. Try this experiment to see what is happening.

You'll need:
some food colouring
2 glasses half full of really hot water
2 glasses half full of ice-cold water

1. Add the food colouring to the glasses of hot water. Now slowly pour some of this water into a glass of cold water. What happens to the hot coloured water? Does it float on top of cold water or does it sink? Is it lighter or heavier than cold water?
2. Do the experiment again. This time, pour a glass of ice-cold water on top of the coloured hot water. What happens?

Just as warm air rises above cool air, making the second floor of your school warmer than the first, warm water rises above cool water, making the top of the lake warmer than the bottom. Warm water is lighter or less dense than cold water, and so up it goes.

Every winter as the cold wind blows over the lake, it chills the surface water, which sinks. The warmer water below rises, gets cooled by the cold air, then sinks too. This continual up-down movement of water is called overturn in a lake and upwelling in the ocean. In this way, food and oxygen are constantly being turned over in the water.

31

Fresh and salty layers

When rivers flow into the ocean, the fresh water floats above the salt water. You can see this wherever muddy rivers empty into the sea. Why are there two layers of water? Try this experiment in the sink or over a big wide container.

You'll need:
2 identical juice bottles
salt
food colouring
an index card

1. Fill the bottles half full of water. Stir as much salt as possible into one bottle — make it really salty — and add some food colouring.
2. Put the index card over the other bottle to make a lid. Now invert the fresh-water bottle over the salt-water bottle.
3. Take away the index card. Does the fresh water float or sink?

Fresh water is lighter or less dense than salt water and so it rises above salt water in the ocean.

Science teasers

Answers on page 103.

1. Water Your Garden Contest
   When is the best time to water your garden in warm weather — during the day or in the early morning or evening?

2. The Ice Trick
   How can you pick up an ice cube using only a piece of string and a salt shaker?

3. The Buoyancy Puzzle
   Is it easier to lift someone up on land or in a pool? Why?

4. Floaters and Sinkers
   How can you make the same piece of Play-Doh sink and then float?

5. Help Me Measure!
   You are making a recipe that calls for 125 mL (½ cup) of margarine. How can you be sure the margarine you scoop out of the container measures exactly 125 mL (½ cup)?

6. The Ice-Cube Puzzle
   When this ice cube melts, what will happen to the water in the glass? Will the water spill over or stay at the same level in the glass?

3. Water in the world

Our planet is called Earth, but it should have been called Water. More than 70% of the Earth's surface is covered by a blue-green layer of ocean. Our ancestors didn't realize how much water there was. They thought that the world was mostly dry land. Little did they know that all the dry land in the world could fit into the largest ocean, the Pacific!

Long ago, people invented myths and stories to explain how the world was formed, why people were created and where the water and land came from. The

Hurons and Iroquois in North America told this story to explain how the earth we live on came to float on the back of a giant turtle swimming in the ocean.

Long, long ago, when the world was young, there were no people. In the beginning there was only water, which stretched as far as you could see. Here lived many animals, such as otters, muskrats and turtles. In the sky, far above, lived the great warriors who hunted in the day and slept in long huts at night. There was also a great tree that grew in the heavens, a tree of life whose four white roots stretched in the four sacred directions.

One day, the ancient chief pulled up the tree, leaving a huge hole through which the hunters could see the waters below. The chief's wife came too close to the enormous hole and fell through into the waters below, clutching a handful of seeds as she fell.

The creatures below saw the woman falling towards them and decided to help her. Two birds flew up to catch her in their wings and held her aloft while the different water animals, one by one, dove down into the deep water to try to find some earth. Finally the muskrat swam all the way to the bottom, scooped up a handful of earth and placed it on the turtle's back. The turtle's shell grew larger and larger and formed the land. The two birds brought the divine woman to the earth. The seeds fell out of her hand and plants grew.

This is how life began on Earth and how dry land came to float on the back of a turtle that was swimming in the ocean.

The great flood

In many parts of the ancient world, people also told stories about a great flood that once covered the entire earth. The Inuit described a terrible flood that killed almost all their ancestors. Only those who lashed their boats together to make a raft managed to escape. Finally, a sorcerer commanded the wind to be still, and the flood waters went down. The Incas of South America also told a story about a great flood that destroyed the world. It was caused by the god who had created the Incas.

The Hebrews who lived in ancient Palestine told of a dreadful flood that lasted 40 days and 40 nights. This famous story is narrated in Genesis, the first book of the Bible. Noah and his family were warned by God to build an ark to save themselves and a pair of each kind of animal. The waters rose and covered the land, but Noah and his family and the animals survived.

Up until recently, many scientists believed that Noah's story described a devastating flood that actually happened near the ancient city of Ur, located in the desert near the modern city of Baghdad, Iraq. It was thought that the two great rivers in the region, the Tigris and the Euphrates, overflowed their banks because of an earthquake and a cyclone, and the water covered the land all around.

But now some geologists believe that much of the earth was flooded more than 11 600 years ago, which explains why so many myths around the world describe devastating floods. The worldwide deluge was caused by the melting of a huge ice cap that covered North America during the last Ice Age. The water poured into the Gulf of Mexico and caused the oceans to rise around the world, flooding the lands and forcing people to move inland.

Why is the sea salty?

According to a Danish legend, two female giants are responsible for all the saltiness of ocean water. A Scandinavian king who captured them from the land of the giants ordered them to grind salt from two magic stones. They ground so much salt that they sunk the ship they were on. Today, they sit at the bottom of the ocean, still grinding salt.

Divine rulers of the water kingdom

Water was a powerful force to ancient people. During storms, the heavens opened and poured down torrents of water. Towering waves sank ships and flooded coastal towns. What god could be powerful enough to control the water? For the Greeks, this super god was called Poseidon; the Romans named him Neptune.

Neptune was a huge fierce god with a long beard. He is often pictured like this, holding a trident, a three-pronged spear. If he became angry, he would wave his trident to bring forth a storm, shatter rocks and shake the shores. But when he rode in his chariot over the sea, the waters became calm again and the fierce monsters were tamed. No one dared harm his favourite pet, the dolphin.

The Inuit of North America and Greenland believed that a huge one-eyed goddess called Sedna controlled the sea and its animals. She did not like humans, and they feared her more than any other divinity because she ruled the seals and whales that they hunted for food. Fortunately, other good spirits such as Agloolik, who lived under the ice, and Nootaikok, who ruled the icebergs, helped the Inuit hunters find seals.

In some ancient cultures, people thought that the lakes and rivers, too, were ruled by powerful supernatural forces. In the year 22 A.D., according to one Japanese legend, the river god appeared to the emperor in a dream and advised him to sacrifice two men to stop a flood. One man was sacrificed while the other man escaped.

If you had grown up in ancient Japan, your parents would have warned you to watch out for Kappa, a dwarf who pulled innocent victims into the water, causing them to drown. Luckily, you could outsmart Kappa by bowing down low to him. To be polite, he had to return the bow, which made all the water pour out from a hole in his head. Once this happened, he could no longer drown anyone.

If you lived among the ancient Slavs in Russia, you would have feared a dangerous spirit called Vodyanoi ("voda" means water). He and other water spirits lived in an undersea palace made of crystal and decorated with gold from sunken ships. At night they would search for unsuspecting swimmers to serve as slaves in their underwater kingdom.

Without rain, nothing can live, and so there were many beliefs and rites about rain and where it came from. The rain-maker was a very important person in many communities all over the world. Ancient Greek rain-makers put a branch of the sacred oak tree into water to produce rain and prayed to Neptune's brother, the all-powerful Zeus, who controlled rain. To the Hebrews, rain was a blessing from God for obeying the law. Drought was a punishment for disobeying him. And so it was that bathing in holy waters came to mean that people were purified of their wrongdoings in many religions.

For Hindus, one of the holiest things you can do is bathe in the River Ganges to wash away your sins. This sacred river, which starts high up in the Himalayas, is the form taken by the goddess Ganga, and is considered the source of all life.

Monsters of the deep

Imagine that you are sailing on unknown waters. You've been warned to watch out for gigantic sea monsters. One night you spot a huge serpent surfacing beside the boat. It's twice as long as the boat, and it's breathing fire. Sound crazy? It wouldn't to the sailors who swore they saw these strange creatures rise out of the depths hundreds of years ago.

- European sailors feared Balena, which had a lion's teeth and claws that ripped apart ships. Near Iceland lived other monsters the size of mountains that could overturn ships. One monster that filled sailors with dread was the kraken, whose arms could reach up and pluck men off the rigging of a ship. Some of these gigantic monsters might have been whales, but the kraken was likely a giant squid.

- Sailors believed in mermaids. These supernatural sea-women dwelled in splendid undersea palaces and could change themselves into humans and come ashore. Spotting a mermaid was considered dangerous because it meant there would soon be a dreadful and deadly storm. Tales of mermaids may have started after sailors saw a dugong or sea-cow, which has a fish's tail but a body that might look human, seen from afar.

- The oldest sea monster, people say, still hides in the dark murky depths of Loch Ness in Scotland ("loch" means lake in Gaelic). Nessie, as she is called, was first seen in 565 A.D. by St. Columba, an Irish missionary. He said that he had scared away "a very odd-looking beastie" when it was about to eat a "pretty lassie" (girl). Since then, many people have reported seeing a monster with a long neck, three humps on its back and a scaly tail. Nessie has been described as a dragon, whale and plesiosaur (an ancient reptile). Photographs of the monster, however, tend to look grainy — Nessie always keeps her distance.

- Creatures similar to Nessie have been spotted in North American lakes. They include the Ogopogo of Lake Okanagan in British Columbia, the monster of Lake Massiwippi in North Hatley, Quebec, and Champ of Lake Champlain in Vermont. Champ was spotted by Samuel de Champlain, the French explorer who named the lake and explored the St. Lawrence River in the 1600s.
The Ogopogo, Canada's most famous water monster, was first seen by Native people who offered it live animals to eat so it wouldn't destroy their canoes.

The real story of water on Earth

Today we know much more about our watery planet. The ancient myths and legends have been replaced by new theories about how water came to cover planet Earth. The modern story goes something like this.

In the beginning, Earth was a red-hot ball of burning liquids and gases, which gradually cooled to form a solid crust of rock. Water appeared thousands of years later. How? Hot gases from the planet's inside core gushed out through volcanoes and geysers and enveloped the hot molten rock in layers of cloud so thick that they blotted out the sun. These gases made up the atmosphere, which acted as a protective blanket for the world below.

Over time, the planet cooled. The water vapour turned into rain. Showers of hot rain poured down for hundreds of years. Enough rain fell to fill the large rock basins in Earth's crust and form one huge ocean, named Panthalassa (from the Greek word meaning "one sea"). There was also one super-continent, called Pangea (meaning "one land"). About 200 million years ago, this ancient continent began to split apart, and the pieces drifted to form our present continents. If you look at a globe, you can see where the continents fit together like jigsaw pieces. Do you see where to place Africa? Can you find a route through the four oceans without touching land?

As the rain fell, it also formed ancient streams and rivers that cut down through the land to form valleys and lakes. And it flowed down through the soil, collecting

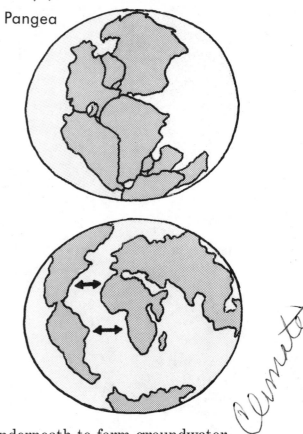

Pangea

underneath to form groundwater.

Over millions of years, the climate on Earth has changed dramatically. Geologists believe there have been five ice ages, periods when much of the surface water froze into thick sheets of ice over parts of the earth and then retreated. At one time, the oceans shrank and there was more dry land at the edges of the continents than there is today. This land, called a continental shelf, is now under water.

The ice sheets over Greenland and the Antarctic are thought to be leftovers from the last Ice Age, which ended about 12 000 years ago. Many of the lakes in Canada and the northern United States were formed at this time. The melting glaciers moved slowly across the land, scraping deep gigantic hollows that gradually filled with water.

Exploring the seas

It took many, many centuries to explore the vast surface of the oceans — and to explore the depths. Only after many hundreds of years went by did explorers show that the world was round, not flat, and that it contained far more water than anyone had ever dreamed possible.

In the 1400s and 1500s, the dream of finding a sea route to the Orient prompted many European kings to send explorers westwards across the unknown waters of the Atlantic.

Christopher Columbus thought that only a small sea separated Europe from Asia. In 1492 he discovered that two continents blocked the way.

Ferdinand Magellan was first to sail around the world, from 1519 to 1522.

The huge size of the Pacific Ocean was discovered by Englishman James Cook, who mapped the way to Australia and the Antarctic in the 1700s.

Early explorers studied the surface of the oceans in their quest for food and travel routes to new lands. Then, about a century ago, undersea explorers began mapping the oceans' depths. In fact, oceanographers (scientists who specialize in the study of the seas) have discovered more about the ocean's secrets in the last 40 years than in all of history. One extraordinary finding is that the world's longest mountain range, called the Mid-Ocean Ridge, extends for 64 000 km (40 000 miles) all around the world. Hot lava oozes out from cracks, or rifts, in the valleys, making a brand-new sea floor.

The most recent explorers, the astronauts, have provided us with stunning photos of the ocean from space. In these pictures, Earth looks like a blue and white ball covered with smaller patches of green and brown. The blue is the oceans, the white is the clouds and masses of ice, and the green and brown patches are the continents. During the *Apollo 11* flight in 1969, astronaut Neil Armstrong said Earth "looks like a beautiful jewel in space."

Space exploration has also revealed that Earth is the only planet in the solar system that has oceans. Earth's location 150 million kilometres (93 million miles) from the sun means that it has the right temperature range to have liquid water — between 0º and 100ºC (32º and 212ºF). Liquid water is very rare in the universe: on the scorching hot planets closest to the sun, water is found only as a gas, if at all; on the cold planets far from the sun, water exists only in the form of ice.

The oceans

There are three huge oceans in the world — the Pacific, the Atlantic and the Indian — and one smaller one, the Arctic.

To see roughly how much ocean water there is on the globe, draw a circle on a piece of paper, and use a pencil and ruler to cut it into quarters. Colour one of the quarter circles brown and the rest blue. The brown part represents all the seven continents put together, and the blue area represents all the four oceans. All seven continents would fit into the largest ocean, the Pacific. How large is the Pacific? It contains 46% of the earth's water.

The world's water supply

If you poured all the water on Earth into a huge glass, here's what you would have. There's not much fresh water, is there? Only 3% of all Earth's water is fresh. This is what we depend on for our daily needs. Most fresh water is locked up in the polar ice caps in the Arctic and Antarctic. If all this ice were to melt, the oceans would rise more than 40 m (130 feet), as high as a 13-storey building!

We Earthlings depend on a tiny fraction of the total amount of water on Earth — less than 1%. Much of this drinkable water is hidden underground in different layers of rock. A smaller amount lies in the lakes and rivers. The rest is floating in the atmosphere as a gas.

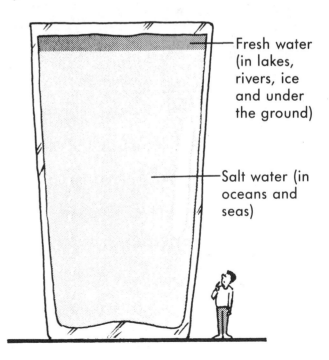

Fresh water (in lakes, rivers, ice and under the ground)

Salt water (in oceans and seas)

Water's endless cycle

Want a glass of fresh water? Fresh, you said? The liquid you're about to drink might be free of salt, but it's hardly fresh. It's an antique. You might be drinking water molecules people drank 2000 years ago! The same water has been going around and around in a cycle for millions of years. Follow water on its round trip around the earth.

Some rain or snow falls on mountains and flows down rivers back into the ocean.

Trees and plants release water vapour, too.

A drop of water that evaporates from the ocean's surface might spend 12 days passing through the air, remain frozen in a glacier for 40 years, lie in a lake for 100 years, or get trapped in the ground for 200 to 10 000 years. (The deeper it gets trapped, the longer it's stuck there.) Eventually the drop will return to the ocean, where it might remain for hundreds of years before starting its journey once again.

Water is used over and over!

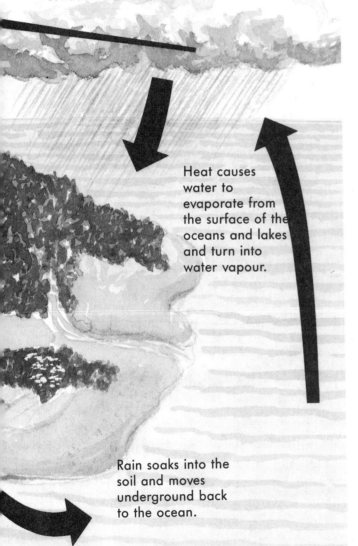

Water droplets form clouds that travel great distances. The water vapour cools and condenses into rain.

Heat causes water to evaporate from the surface of the oceans and lakes and turn into water vapour.

Rain soaks into the soil and moves underground back to the ocean.

Plant breath

You can see a plant's watery "breath" by trying this. Cover a house plant with a clear plastic bag overnight. In the morning, you'll see drops of water on the inside surface of the bag. Some of this water evaporated from the leaves. The rest is condensed water vapour from the air in the bag.

If you're going away for a few days, make plastic-bag tents around your plants. Punch a few holes for air vents at the top of each tent. Give the plants a good soaking before you bag them and they will water themselves!

Water in the air

Every time you breathe, you breathe in water. Water is in the air all around you. It's a small but important part of Earth's water supply. It gives us the rain that makes life possible.

How does water get into the air? Most of it evaporates from the oceans. In a single year, about 319 000 km³ (76 650 cubic miles) of sea water evaporate from the oceans to form rain. One-third of this rain falls on land; the rest falls back into the oceans.

Water vapour also enters the air by evaporating from lakes and rivers, from plants and trees, even from humans — from our watery sweat and breath. It rises into the air and condenses into tiny droplets of water or freezes into ice crystals to form clouds. Winds blow the clouds over areas where the air is cooler. The droplets condense further on particles of dust, become too heavy to float in air currents and fall as rain or snow.

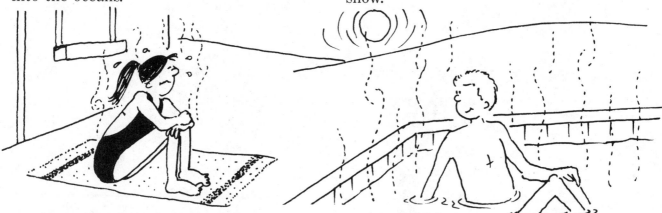

What is it?

Water and weather go together. Test your weather know-how by answering the questions below. Answers on page 103.

1. These can be as small as Smarties or as big as tennis balls, depending on how much they've travelled up and down in cold masses of air. They start off as frozen raindrops.
2. On a below-freezing night, wet dew on the grass freezes into this substance. It coats the plants.
3. As these tiny ice crystals travel down to Earth, they collect dust particles and more ice crystals and grow larger, sometimes into big flakes.
4. This is cloudy air at ground level. It forms when a warm moist air mass is cooled by the land.

SALT WATER

Have you ever wondered what makes the ocean blue or the tides rise and fall? Strap on a life preserver and read on.

Why is the ocean salty?

Salts and minerals from rocks have become dissolved in the ocean, making it salty. This happens gradually, over thousands of years. First the rocks are broken up and ground down into soil by the action of ice, wind and rain. Eventually, rivers carry the soil to the ocean. The salts and minerals don't evaporate as water does; they stay in the ocean. Scientists believe the first ocean wasn't salty at all — the salt has been washed in by rivers over millions of years.

To taste water as salty as the ocean, mix a little less than 5 mL (1 teaspoon) of table salt into a small drinking glass of warm water. Ocean water contains about 1.6 kg (3.5 pounds) of salt to every 45.6 kg (100 pounds) of water. Beware: real sea water tastes far worse! That's because it contains different types of salts besides table salt, as well as nutrients and traces of all the other elements on Earth, including iron and gold.

Ocean water is too salty for humans to drink. How do animals and fish survive in it? They have ways of getting rid of the excess salt. Sea turtles, for example, ooze salty water out in tears, while birds pass the water out holes in their beaks.

Why is the ocean blue?

Sunlight is made up of different colours of light; each colour has a different wavelength. (You can see these colours in a rainbow because the light rays are scattered by the water droplets.)

Blue and green wavelengths of light reach farther down in the water than other colours before they are absorbed. They are then reflected back to the surface, making the water look blue-green. The surface of the water also reflects the sky, so that on clear sunny days the water looks blue, and on stormy days it appears grey.

Water usually looks bluest when it's deep and contains little life. On the coastline of hot tropical areas, tiny plants floating on the ocean's surface turn it green. Still other micro-organisms make the water look red, like in the Red Sea. Rivers appear brown when they flow across a flat plain to the ocean, because of all the mud and silt they're carrying.

What's the bottom of the ocean like?

Amazing, absolutely amazing. It isn't flat, as people believed until only recently. There are mountain ranges higher than the ones on land, deep valleys and canyons, sandy plains and muddy and stony areas.

The Bermuda Islands sit on top of one mountain range. If you walked from the east coast of North America to Europe, this is what the underwater landscape would look like.

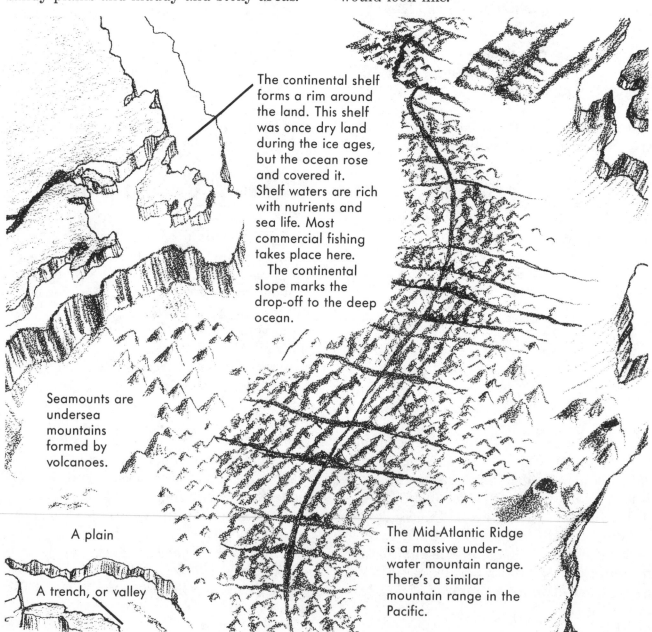

The continental shelf forms a rim around the land. This shelf was once dry land during the ice ages, but the ocean rose and covered it. Shelf waters are rich with nutrients and sea life. Most commercial fishing takes place here.

The continental slope marks the drop-off to the deep ocean.

Seamounts are undersea mountains formed by volcanoes.

A plain

A trench, or valley

The Mid-Atlantic Ridge is a massive underwater mountain range. There's a similar mountain range in the Pacific.

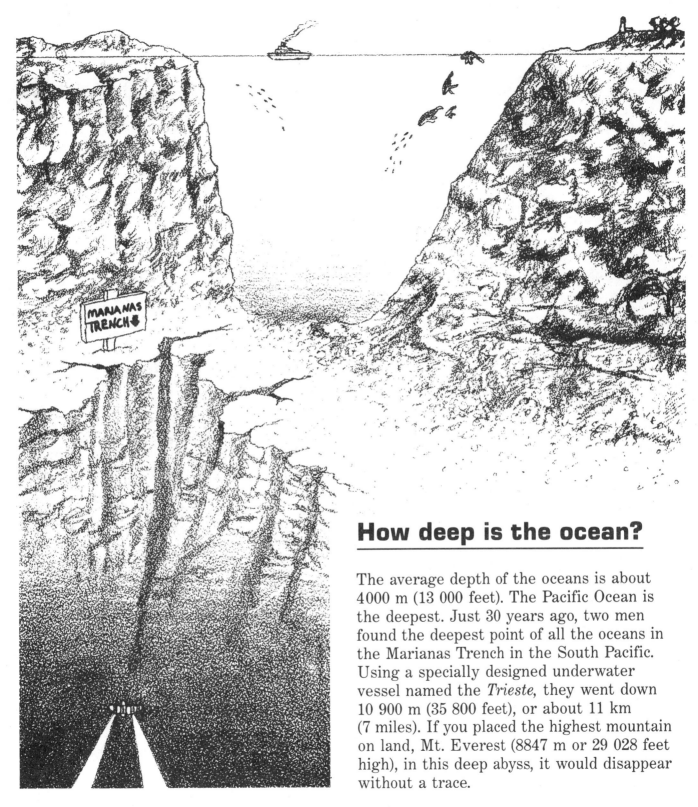

How deep is the ocean?

The average depth of the oceans is about 4000 m (13 000 feet). The Pacific Ocean is the deepest. Just 30 years ago, two men found the deepest point of all the oceans in the Marianas Trench in the South Pacific. Using a specially designed underwater vessel named the *Trieste*, they went down 10 900 m (35 800 feet), or about 11 km (7 miles). If you placed the highest mountain on land, Mt. Everest (8847 m or 29 028 feet high), in this deep abyss, it would disappear without a trace.

I'm down so low and feel so blue...

What does the watery world look like to SCUBA divers? The red and ultraviolet wavelengths of sunlight get absorbed first by the water, which means that SCUBA divers can no longer see reds deeper than 6 m (20 feet). As divers go deeper, red objects look muddy brown, and yellow things become pale yellowish green. Below 18 m (60 feet) everything looks blue-green. SCUBA divers who accidentally cut themselves can get quite a nasty shock when they see green blood oozing out. The blue-green world becomes grayer, then blacker, the deeper they dive.

SCUBA divers report other surprises, too: from down below, rainwater looks like an oily film as it spreads over the water.

What makes the waves?

The winds. Most ocean waves are less than 3.5 m (12 feet) high. The stronger the wind and the greater the distance the wind blows over the water, the higher the waves. The highest wave ever recorded was measured in 1933 by an American tanker in the Pacific. It was 34 m (112 feet) high, as tall as a nine-storey building. Immense waves are sometimes created when volcanoes erupt under the sea. (See "Weird water," page 70.)

What are tides and what causes them?

Winds push waves along the top surface of the ocean, but the tides move the entire ocean. At one time, people thought the daily rise and fall of the ocean meant that it was breathing. Now we know that tides are caused by the gravitational pull between Earth, the moon and the sun.

The moon's gravity acts like a powerful magnet that makes the water rise up into a bulge on the side of Earth that faces the moon. When the bulge is far away from land, it's low tide. When the bulge meets the land, it's high tide. The pull is greatest twice a month during the full moon and new moon because the sun and moon are in line with the earth, and so act together as a team. This causes very high "spring" tides. "Neap" (low) tides occur when the sun and moon are at right angles to one another and tend to cancel each other's gravitational pull.

Tides vary greatly in different parts of the world. The highest tide occurs in the Bay of Fundy, Nova Scotia, due to the funnel shape of this long bay. The difference between the low and high tides is 14 m (45 feet). This allows fishermen to pick the fish out of nets after the tide goes out.

Tides flush inlets and channels with new sea water, preventing the water from becoming stagnant. They wear away shorelines but also create new land as they deposit sand and silt. As you read this book, the world's tides are shifting enough rock and sand to build a mountain.

What are currents?

Think of them as moving highways of water travelling across the ocean's surface or deep below. They carry huge amounts of water around the world and in this way help to circulate heat, nutrients and oxygen essential to sea creatures. They make ships travel faster or slower, so sailors have to know how to navigate along these highways.

Surface currents are driven by winds. Warm water moves away from the equator and cooler water from the poles moves in to replace it. Because of the earth's spin, currents circle to the right in the Northern Hemisphere and to the left in the Southern Hemisphere.

One of the greatest currents is the Gulf Stream, which is 65 km (40 miles) wide and 3220 km (2000 miles) long. In the winter this ocean highway carries warmth from the Gulf of Mexico up the coast of the United States, then veers across to Europe. Thanks to the Gulf Stream, England enjoys warmer winters than Labrador, which is at the same latitude.

Deep ocean currents are caused by differences in the water's temperature and salt content — a river of icy-cold salty water sinks under a river of warmer, less salty water. When these cold-water currents rise up, they "feed" the surface with rich nutrients and oxygen, creating excellent fishing areas, such as the Grand Banks off the coast of Newfoundland. For more about currents, see "Hot and cold layers," page 31.

Are there living creatures everywhere in the ocean?

Yes, even in the deepest, coldest and blackest parts. In general, the deeper you go, the fewer living creatures there are. Most sea life is found in the upper 100 m (300 feet) of water because this is as far as sunlight can penetrate to support the plants many sea creatures feed on.

The ocean is divided into zones in which a variety of life exists. Each area has its own ecological system (ecosystem), a community of plants and animals adapted to the conditions in that area. There are many different marine ecosystems — from saltwater marshes and rocky beaches to coral reefs and the deep ocean.

tidepool

1. The intertidal zone is where the land and sea meet. This is a harsh place to live. When the tide comes in, the animals and plants live underwater; when the tide goes out, they must somehow survive on the rocks, sand or mud, or in small tidepools. Tidepools are wonderful places to explore for mussels, crabs and sea stars.

2. The continental shelf waters lie between the shoreline and the open sea. They form a zone of shallow ocean over the continental shelf. This is the home of most sea animals.

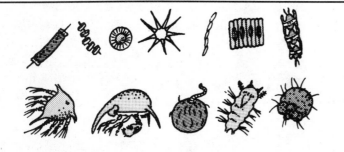

Trillions of tiny plants and animals called ''plankton'' are the food for sea animals. Plant plankton produce most of the oxygen we breathe. There may be 12 million tiny plankton in a litre (about 1½ pints) of sea water.

Zooplankton are tiny animals the size of the letters in this sentence. They eat plant plankton and get eaten by larger animals.

What does it sound like under the sea?

It's surprisingly noisy. Volcanoes hiss, rain patters down and sea creatures snap, squeak and croak as they search for food or flee from their enemies. During World War II, submarine crews were startled by strange crackling sounds, which turned out to be snapping shrimps. (Snap your fingers — this is the noise these shrimps make.)

The noisiest creatures are the mammals.

Whales send out different calls when they court one another, sense danger or want to keep the pod (group) together. Whale sounds can travel hundreds of kilometres (miles) underwater. Look for a record of whale "songs" in the library. Porpoises, too, have a highly developed communication system and make clicking noises to "talk" to one another.

3. The open ocean, or sunlit zone, is home to fewer creatures. Fish and animals here are dark on top, making them hard to see when enemies look down into the water. They are light underneath, making them hard to spot by enemies swimming below.

4. The zone of darkness is pitch black and cold. Deep-sea creatures eat one another or feed on dead creatures. Fish have huge mouths that can open wide enough to swallow fish larger than themselves; their stomachs are stretchable. Some have light-producing organs. The female deep-sea angler fish, for example, lures animals into her mouth with a unique "fishing rod" attached to her head — the rod lights up.

5. On the sea bottom live many animals, such as burrowing worms, sea cucumbers and starfish. They eat dead and decaying animals and plants that drift down from above.

55

Which sea monster am I?

Many sea creatures have been called monsters because of their huge size or strange appearance. Can you match these creatures to the clues on the opposite page? Answers on page 103.

blue whale

shark

oarfish

manta ray

dugongs

coelacanth

giant octopus

giant squid

1. I'm a cold-blooded animal with eyes as big as dinner plates. I can grow longer than your classroom. Humans discovered we're real after one of us attacked a fishing boat off the Newfoundland coast in 1873. The fisherman's 12-year-old son, Tom, hacked off a tentacle with a hatchet. Who am I?

2. They call me the wolf of the sea because I'll eat anything. I don't eat many humans, though, despite what movies say about my jaws and sharp teeth. Who am I?

3. I'm the largest animal in the entire world and may grow longer than two buses parked end to end. I eat small shrimp plankton called krill, which I strain through my mouth. Who am I?

4. Do I ever scare fishermen when I leap out of the water and sail through the air like a giant bird! Who am I?

5. People used to think my family died out 70 million years ago! Because of their stumpy fins, my Stone-Age ancestors are believed to be among the first fish that walked on land. Who am I?

6. I'm shy unless someone disturbs me in my cave. I can change colour in a flash or squirt out inky clouds so that enemies can't smell or see me. Who am I?

7. I've been called a giant sea serpent. My dorsal fin looks like a bright crimson mane, and my pectoral fins look like oars. I have a neat trick for escaping sharks — I let them bite off my tail and I swim away! Who am I?

8. Sailors thought we were mermaids. I use my flippers to cradle my baby. My face looks human when seen from afar. No one has ever caught us alive because we're so very, very shy. Who am I?

FRESH WATER

Which of these is fresh water?

a. non-salty water
b. fast-flowing water in streams
c. stagnant water in swamps and ponds
d. cold tap water
e. rainwater

Congratulations if you said all of the above. Fresh water is found under the ground, in rivers and lakes and reservoirs, in icebergs and ice caps, and in rain. This is the water you use for drinking and washing, the water that's dammed to make electricity, and the water factories use to make thousands of products — the bread you eat, the cars you ride in, the paper this book is printed on.

There's lots of fresh water on Earth, but three-quarters of it is frozen at the North and South poles. Less than one-quarter of the world's total fresh-water supply is found in rivers and lakes or underground. In some places, underground water, or "groundwater" as it is called, is buried so deep that it's too costly to pump up. In other places there is just not enough rain for a lot of groundwater to collect. This makes fresh water very precious. It's like liquid gold.

If you live in the country, your water likely comes up pipes connected to an underground well or a nearby spring. The well draws upon groundwater that has collected in the loose gravel and soil above a layer of rock the water can't get through. Groundwater is usually almost entirely free of bacteria, which makes it an attractive source of water for many farms, villages and towns — if it's not polluted. More than half the people on Earth depend on groundwater for their water supply. In some countries, such as Malta and Australia, groundwater is the *only* source of water for most of the population. Look at a map to see why: Malta is an island too small to receive much rain, while most of Australia is a desert.

A huge area of loose sand and gravel filled with groundwater is called an aquifer, and the top level is called a water table. Dig anywhere in the world and you'll eventually find water, even under the Sahara!

The water table in an aquifer can fall dangerously low during a hot spell or if people and industries are drawing upon the supply faster than nature can replace it. So much water has been pumped from the aquifer under Houston, Texas, that the city has sunk a couple of metres (several feet)! Overpumping has also led to the land sinking under Mexico City and Venice, Italy. In some areas, such as southeast Asia, cities along the entire coastline are sinking, centimetre by centimetre, into the sea as thousands of wells drain the water from the clay beds under the cities.

The Ogallala aquifer in the southwestern United States is the largest underground body of water in the world. It was formed at the end of the ice ages, thousands of years ago. People who live on top of the Ogallala aquifer are concerned that too much water is being pumped out to irrigate crops and raise cattle. Some hydrogeologists (scientists specializing in water and geology) estimate that this ancient water source may dry up in less than 30 years. Only one-tenth of the aquifer is replaced by rainfall each year.

The irrigated farmland over this aquifer produces much of the beef and grain for the United States. Worried about the possible shortage of water, some scientists have drawn up plans to pump water from the Great Lakes to this region, much to the anger of the leaders of the states and provinces who live around these lakes. This is one example of how water may become the cause of political fights between water-rich and water-poor areas in the future.

Besides overpumping, groundwater faces another enemy — pollution. Unfortunately, groundwater can be easily polluted. The soil acts like a giant sponge that soaks up the things we dump — buried chemical waste from factories, garbage from landfill sites, lead from the paint you throw out, insecticides from fields and even bug killers from gardens. Gas stations store fuel in underground steel tanks that rust, and the gas leaks out. It only takes one litre (quart) of gasoline to pollute up to a million litres (quarts) of groundwater.

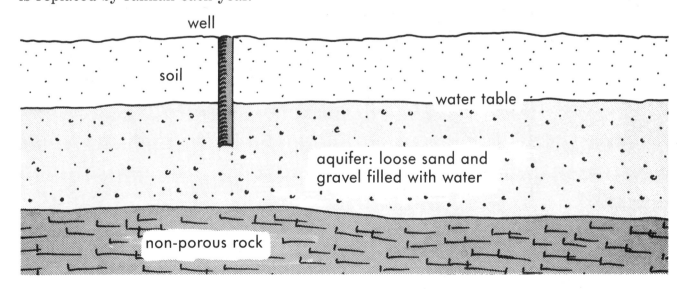

59

Make your own mini-well

Here's how to make a well that draws water from a small aquifer.

You'll need:
a large glass container or a cardboard
  box lined with plastic wrap
some sand
a jug of water
small stones

1. Fill the container three-quarters full of sand.
2. Slowly pour in enough water to make the sand damp.
3. Dig a small hole with your finger in the centre of the sand and line it with small stones. This is your well. Watch it fill with water to the level of water in the sand.

If you keep adding water to the sand every day, the sand will stay moist and your well will continue to fill with water. What happens when you stop adding water for a few days and put the container in a warm, dry place? You've created a drought: the water evaporates and your well runs dry.

- There's enough water in North American aquifers to cover the continent in 30 m (100 feet) of water.
- Aquifers near the surface of the ground produce lakes and rivers; aquifers on mountains come out as springs.

- Some people, called "dowsers," say they can locate underground water with the aid of a divining rod, a branch with a fork in it. They hold the branch in both hands as they walk across the land. When the forked end bends down, dowsers say there's water underneath.

60

Rivers

Suppose you could time-travel back to the year 1840. You and your family arrive in North America on a ship that docks in New York City. Another boat takes you up the Hudson, through the locks of the Erie Canal to Buffalo, and across Lake Erie to a small village in Ohio. From there, your family travels south to look for some land to clear for a farm. Finally your parents spot the land they want — a piece of forest that slopes down to a small river.

"I thank my lucky stars," your father says with a grin. He's happy he won't have to dig a deep well. The water for the house, barn and fields can come from the river.

"Great! We'll have trout tonight," says your mother happily as she heads to the river bank with her fishing pole.

"Yippee. Time for a swim," you shout, running after your mom.

Ever since people settled down in one place to grow crops, they have built towns and villages along rivers, lakes or near river mouths. Rivers provided clean drinking water, plenty of food and water for farming. In some places, rivers protected townspeople from enemy attacks and provided a highway to travel along.

Later, towns grew up at places where bridges could be built across a river, or where rough water forced boats to unload. You can sometimes tell that a city is located beside a river because its name includes the word "bridge" or "rapids" or "ford." (A ford is a shallow place to cross.) Can you name the U.S. President who grew up in Grand Rapids, Michigan? (Hint: his name appears in this paragraph.)

Many factories were located on rivers, too. The factories needed the water to make their products, cool down their machines and dump their waste. Fast-moving rivers, especially those that flow over waterfalls, were also used as power sources to turn water wheels in mills and grind cereal into flour.

Today water-power is harnessed to make electricity. Power produced by water is called hydroelectric power ("hydro" means water in Greek).

How river smart are you?

Try matching these river names to the correct clues on the next page. (Need some help? Use a world atlas!) Answers on page 103.

1. Ganges
2. Nile
3. Amazon
4. St. Lawrence
5. Mississippi
6. Thames
7. Colorado
8. Niagara
9. Rhine
10. Columbia
11. Hwang Ho

a. The great waterway of Europe, and one of the busiest rivers in the world. It starts in the Swiss Alps and ends in Rotterdam, Holland.

b. Also called Old Man River, this is the largest water highway in the United States. Its name is fun to spell, but its water is muddy because of all the silt it carries to the Gulf of Mexico.

c. Fed by more than a thousand smaller rivers, this mighty river discharges enough water into the ocean to fill Lake Ontario in three hours. It sustains the world's largest rainforest, contains one-fifth of the world's fresh water and travels through half of South America.

d. This river is worshipped by Hindus who come to bathe in its holy waters. It starts as an icy river in the Himalayas and becomes very polluted near cities.

e. Over millions of years, this river has carved a great canyon, one of the most spectacular sights in the world. Its name in Spanish means red because of the red sediment it carries.

f. Large ocean-going ships can travel more than 3200 km (2000 miles) from the Atlantic to ports on the Great Lakes if they follow this river. Recent studies have shown it is terribly polluted.

g. Known also as the Yellow River and "China's Sorrow," this river brings death and disasters when it floods.

h. One of the great rivers of the ancient world. Its yearly floods created rich farmland for peasants. A huge dam near its mouth now provides electricity to much of Egypt.

i. The most important river in Britain. It once was so polluted that many people died from drinking its water, and it smelled so bad that politicians couldn't meet in the House of Parliament located beside it. In the 1950s, it was declared a dead river because of the sewage. Efforts to clean it up have been successful: fish now swim in it again.

j. Over the years, this powerful river has cut away the limestone cliff over which it falls. This famous waterfall has now moved 11 km (7 miles) farther back along the river. Scientists think that after thousands of years, the falls will recede all the way to Lake Erie!

k. Lewis and Clark canoed down this river to the Pacific Ocean. Its name can be traced to another explorer, however, even though he never came close to it. Today this river defines the boundary between Washington and Oregon.

An airplane view of water

One of the best ways to appreciate how much water there is in much of the United States and Canada is from an airplane. Suddenly you can see how long the rivers are, how enormous the Great Lakes are and how many smaller lakes exist. As you land, you can also go nuts trying to count all the swimming pools!

Why are there so many lakes? About 30 000 years ago, during the last Ice Age, most of Canada and parts of the United States were covered in a massive sheet of ice. As the ice moved south and west, it carried with it boulders that acted like gigantic spoons that scooped humungous holes in the rock. The ice melted and filled the holes. The result: more lakes than you can count!

Imagine you're in a plane flying over these bodies of water. Look at the shapes below. Can you match up these clues to the correct drawing?

a. lake created by a dam
b. small lakes along a river
c. swimming pool

Answers on page 103.

Anyone for a swim?

Lakes provide drinking water, resting places for migrating birds and homes for many types of water dwellers and fish. They're also great places to go boating and swimming. Slip on your swimming suit and test out these unusual lakes.

- Put on SCUBA diving gear if you want to explore Lake Baikal in southern Siberia. It's the world's deepest lake. It may be the world's oldest lake, too, since scientists believe it's more than 65 million years old and was once part of the Arctic Ocean! It is home to many plants and animals found nowhere else, including fresh-water seals. In 1990, more than 100 scientists began exploring it, using underwater robots and submersibles.

- Even poor swimmers can float easily in the Dead Sea in Israel. Why? It's seven times saltier than the ocean, making swimmers super-buoyant. The Dead Sea is really a lake with no outlet. What makes it so salty? In the hot climate the water evaporates, leaving a load of salt behind. It's called "dead" because fish can't live in its super-salty waters.

- Marathon swimmers might want to swim the length of the world's largest lake, Lake Superior. It's about 640 km (400 miles) long. Sometimes its waves are so high you might think you're in the ocean. The waves were no problem for Canada's Vicki Keith, who managed to swim all the Great Lakes, including Lake Superior, in 1988.

Exploring ponds and lakes

The best way to learn about life in fresh water is by exploring a pond or lake. You'll need to be patient, though, and move around silently. When you catch a creature, put it in a container of pond water so that you can look at it more closely. Put specimens back in the water when you're done.

You'll need:
a pad of paper and pencil
a nature guide to ponds (optional)
a small hand-dip net or a flat-bottomed net, and a waterscope (see "Making your tools" box)
2 or 3 clear containers half full of pond water
a spoon
a kitchen sieve
a magnifying glass

Making your tools

A hand-dip net is good for collecting small insects and organisms on top of the water and among plants. Bend a coat hanger into a ring about 8 cm (3 inches) in diameter. Sew a nylon stocking or piece of cheesecloth onto the ring. Fasten the ring to a stick or pole. A butterfly net will work fine, too.

A flat-bottomed net is useful for collecting fish, minnows or beetles. Bend a coat hanger into a D-shaped frame about 30 cm (a foot) wide. Sew a net bag made from cheesecloth onto the rim. Fasten the ring to a broom handle.

A waterscope is the ideal way to view tiny animals and plants without getting wet! Cut off the bottom of a plastic pail or ice-cream container. Cut a piece of plastic wrap to cover the bottom as shown. Attach the wrap with large elastics. Use your waterscope on a sunny day in clear, still water.

How healthy is the water?

Is your local stream healthy or polluted? If you answer "yes" to any of these questions, it may be polluted. Tell your parents and contact the nearest environmental agency.

- Does the water smell like rotten eggs? If so, it might mean that sewage (waste water from toilets and sinks) is being dumped into the water.
- Is the water brown and muddy? Too much soil in the water makes it difficult for plants to grow and fish to breathe. Check the banks to see if soil is being washed directly into the stream.

- Do you see any patches of oil on the water? Oil may be leaking from an abandoned oil drum or from a factory or motorboats.
- Is the water dark green and full of algae, with no sign of fish or other types of water animals? Algae may be using up all the oxygen in the water and choking off other forms of life.

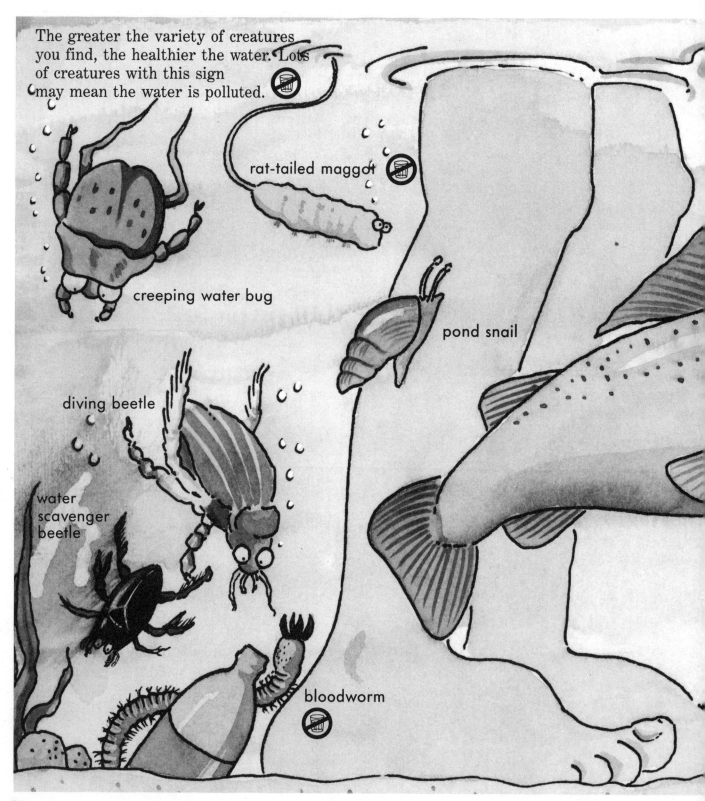

The greater the variety of creatures you find, the healthier the water. Lots of creatures with this sign may mean the water is polluted.

rat-tailed maggot

creeping water bug

pond snail

diving beetle

water scavenger beetle

bloodworm

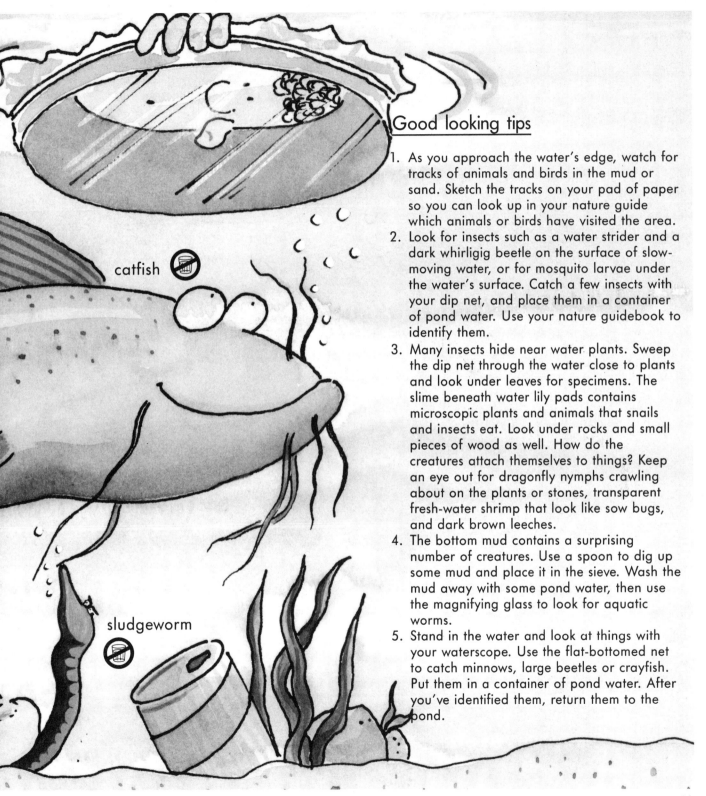

catfish

sludgeworm

Good looking tips

1. As you approach the water's edge, watch for tracks of animals and birds in the mud or sand. Sketch the tracks on your pad of paper so you can look up in your nature guide which animals or birds have visited the area.

2. Look for insects such as a water strider and a dark whirligig beetle on the surface of slow-moving water, or for mosquito larvae under the water's surface. Catch a few insects with your dip net, and place them in a container of pond water. Use your nature guidebook to identify them.

3. Many insects hide near water plants. Sweep the dip net through the water close to plants and look under leaves for specimens. The slime beneath water lily pads contains microscopic plants and animals that snails and insects eat. Look under rocks and small pieces of wood as well. How do the creatures attach themselves to things? Keep an eye out for dragonfly nymphs crawling about on the plants or stones, transparent fresh-water shrimp that look like sow bugs, and dark brown leeches.

4. The bottom mud contains a surprising number of creatures. Use a spoon to dig up some mud and place it in the sieve. Wash the mud away with some pond water, then use the magnifying glass to look for aquatic worms.

5. Stand in the water and look at things with your waterscope. Use the flat-bottomed net to catch minnows, large beetles or crayfish. Put them in a container of pond water. After you've identified them, return them to the pond.

Weird water

Waterspouts

You've seen photos of tornadoes. Now imagine a white twisting column of water rising from the surface of a lake or ocean. These "waterspouts" form over lakes in late summer and over oceans in tropical areas when thunderstorms are about to occur.

Geysers

Jets of water and steam that gush out of the ground in volcanic areas are called geysers. Hot lava deep in the ground heats nearby water so hot that it erupts in a column of steam and water, like a kettle boiling over. A geyser might shoot up as high as a 20-storey building. The world's most famous geyser is called Old Faithful because it erupts about once an hour. It's found in Yellowstone National Park, U.S.A. Other geysers can be seen in New Zealand and Iceland.

Hot spots in the ocean

Up until the 1960s, everyone believed the ocean bottom was inky black and freezing cold. What could possibly survive there? Recent submarine explorations have discovered that hot spots do exist in the ocean, where cracks in the bottom split apart, spewing out lava. Chemicals in the scalding water provide food for bizarre creatures — bright red worms up to 2 m (6 feet) long that live in white tubes, and red-blooded clams as big as dinner plates.

Floods

Never is the power of a river so awesome and frightening as during a flood. Houses and people are swept away by the surging water and farmlands are sometimes ruined. Floods in North America usually occur in the spring when rains or melting snow cause the water in rivers to rise too fast.

Many countries try to control floods by building high banks, called levees, along a river to keep in the water. Some levees on the Mississippi are as high as a two-storey house. Dams are also built to control the water flow. When the water rises in the rivers, extra water is stored in reservoirs behind the dams.

Killer waves in the Pacific

The Japanese call these immense waves "tsunami" (soo-nami). Set in motion by an underwater earthquake, volcano or nuclear blast, they spread out like ripples in a pond, moving incredibly fast and building to enormous heights, up to 67 m (220 feet), as they approach the coastline. In the past, devastating tsunamis have lifted sailors and sharks into treetops, flung boats on mountains and killed thousands of people in minutes. Tsunami warning systems in the Pacific are now set in operation when scientists believe an undersea explosion will cause these killer waves.

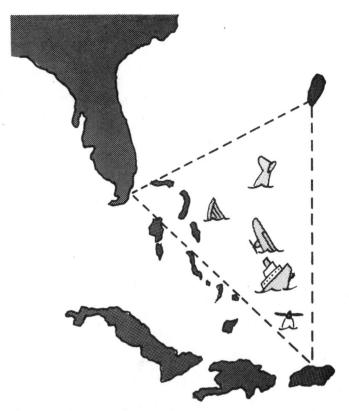

The Bermuda Triangle

More than 150 ships and planes have mysteriously vanished in a triangle of water between Bermuda, Puerto Rico and the Florida Coast. The Bermuda or Devil's Triangle has had a spooky history since claiming the lives of more than a thousand sailors and pilots who have disappeared without a trace in the past 150 years.

What causes the disappearances? No one knows for sure. One recent theory, suggested by Canadian scientist Donald Davidson, holds that giant bubbles of methane gas might be rising from the ocean floor, forming a froth that sinks ships and causes engine failure in planes. Others have proposed that sudden storms and weird weather, geological differences, smugglers — even visitors from outer space — explain the mystery.

71

4. Save our water!

Suppose one day you turned on the tap and only a few drops came out. Sound impossible? Not to people in areas where there has been a drought, a long period with little or no rain. During a drought, plants wither, crops fail, animals die and people go hungry and even starve. The shortage of water produces a shortage of food.

One of the areas affected the most by drought in the world is the Sahel, located south of the Sahara Desert in Africa. More than 30 million people live in the eight countries of this desert-like region. As the population has increased, more food has been needed to feed everyone and more land has been forced to produce crops. The land has been farmed until it has become exhausted, so the desert keeps getting bigger and bigger. And the drought has gone on for several years — many wells have run dry, and large lakes and rivers have shrunk. Without rain, crops fail and people die from lack of food. The countries in the Sahel are now trying to conserve water and build more wells.

In areas such as Canada and Russia there is more than enough fresh water to meet everyone's needs. These areas can count on a constant supply of rain and snow. Yet even water-rich countries face a serious danger to their water — pollution. People in some towns now have to buy bottled water because their water is too polluted to drink.

The population of the world keeps growing. It's expected that the demand for drinking water will double between now and the turn of the century. Many experts believe the world is about to face a water crisis; there won't be enough clean water to go around.

As the demand for water grows around the world, the search is on to find new sources of fresh water:

- Seeding clouds to make them rain. Some scientists have developed ways to force clouds to rain. Planes sprinkle a chemical into the clouds, making the raindrops grow. Only certain types of clouds can be seeded in this way, however — they have to be the right temperature.

- Making sea water drinkable. Water-poor countries located near oceans have developed ways to remove the salt from the ocean's water, a process called desalinization. How? The process copies nature. Huge quantities of salt water are heated so that the water evaporates and leaves the salt behind; the water vapour is cooled so it condenses to form fresh water. Removing the salt from sea water is extremely expensive, however, and demands a great deal of energy. The world's largest desalinization plants are in Saudi Arabia.

- Towing huge icebergs to cities facing water shortages. It would take several months for ships to tow an iceberg from Antarctica to South America or the Middle East, but some scientists believe it would be worth the expense. A half-melted iceberg could provide a large city with fresh water for a few months.

The water-poor versus the water-rich

A person who lives in semi-arid regions in Africa gets by on about 12 L (21 pints) of water a day, less than the amount you send down the toilet with each flush.

Since millions of people in poor countries don't have clean, safe drinking water, the United Nations named the 1980s the Water Decade. Many methods have been used to make more clean water available — installing hand pumps in villages, educating people about killer diseases carried by water, improving the ways to water crops.

- Don't use the toilet as a garbage container. One toilet flush uses 19-27 L (5-7 gallons) of water.

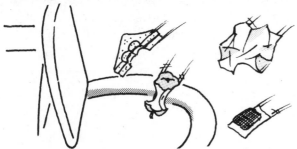

- Use buckets of water to wash the car, not a running hose.

North Americans use more water each day than people in any other part of the world. Why? Most of the water is used on farms to irrigate crops and raise cattle, or in factories to make products requiring tonnes (tons) of water — such as cars and paper — and food, such as bread. The rest goes to our homes where we happily waste it, thinking it's cheap and plentiful.

The average family of four uses more than nine bathtubs full of water a day (1200 L or 316 gallons). How can you save water?

- Turn the tap off when you brush your teeth or wash your hands and save 3 L (5 pints).

- Take short showers rather than baths. A three-minute shower uses half as much water as a full bathtub. Turn off the water when you lather your hair. Suggest that your parents install a new energy-efficient showerhead that reduces the water flow, too.

- Check your house for leaks. How? In some houses you can check the water meter to find out. (If you live in an apartment or if your water meter isn't located where you can read it, go on to the next activity.) Jot down the numbers on the water meter when everyone leaves the house for the day. If the numbers are higher when you return, there's water leaking somewhere.

- Find out if your toilet leaks. With the help of an adult, remove the toilet tank lid and put in a few drops of food colouring. Don't flush the toilet. If the toilet bowl soon after contains coloured water, the toilet tank is leaking. Time for a repair job or a new low-flush toilet.

A water treatment plant

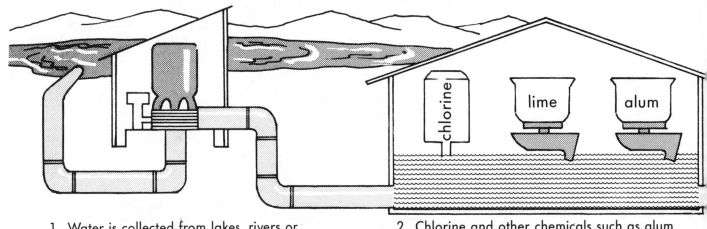

1. Water is collected from lakes, rivers or reservoirs. Large objects are first screened out.

2. Chlorine and other chemicals such as alum or lime are added to the water to destroy impurities and bad taste and smell.

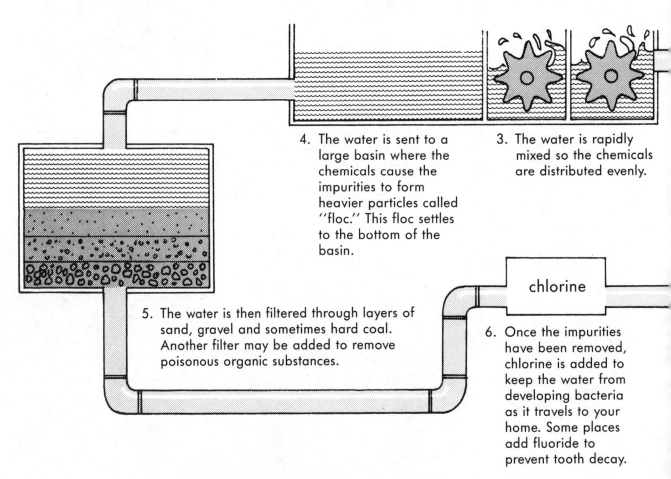

4. The water is sent to a large basin where the chemicals cause the impurities to form heavier particles called "floc." This floc settles to the bottom of the basin.

3. The water is rapidly mixed so the chemicals are distributed evenly.

5. The water is then filtered through layers of sand, gravel and sometimes hard coal. Another filter may be added to remove poisonous organic substances.

6. Once the impurities have been removed, chlorine is added to keep the water from developing bacteria as it travels to your home. Some places add fluoride to prevent tooth decay.

Make your own water filter

Here's one way to clean muddy water.
Warning: Don't drink the filtered water.
It may still contain bacteria.

You'll need:
a paper coffee filter
a funnel
some charcoal or fine gravel
some sand
muddy water
a container

1. Place the coffee filter in the funnel.
2. Put a layer of crushed charcoal or gravel in the bottom of the funnel. Make another layer of sand above it.
3. Pour the muddy water through your filter into the container. What happens to the water? Is it clearer?

In many water treatment plants, dirty water is first allowed to settle in a large basin and then is filtered through layers of sand and gravel. This filtering method copies nature — rainwater seeping through the ground gets cleaned the same way.

7. The clean water is stored in a reservoir or tank and then travels through huge pipes (called water mains) to your house.

What makes the water cycle polluted?

Human and animal wastes are only one source of pollution. A far more serious threat comes from the dumping of chemical wastes into water by factories. Even a small amount of these poisonous chemicals can kill plants and animals that live in or around the water.

Farms are another big source of pollution. Rain washes fertilizers and pesticides (chemicals that kill insects) into the streams and rivers. The fertilizers contain nutrients that make plants grow quickly on the fields. Once these nutrients reach the ponds and streams, they overfeed water, causing algae to grow on the water's surface. The build-up of algae prevents oxygen from reaching the other creatures and they suffocate.

Pesticides can travel thousands of kilometres (miles), carried by the winds. So even though pesticides such as DDT are now outlawed in North America, DDT sprayed in Mexico and South America is carried on the wind and rains into our lakes.

We produce chemical wastes in our homes, too. Every time we flush away chemicals after cleaning, we send polluted water into the sewer system. Stores, restaurants and hospitals also flush chemicals down the sewer. This waste water might contain bleaches and detergents.

Car and factory exhaust releases chemicals that cause acid rain. Salt from roads runs off into streams and rivers, harming plants and animals.

Untreated sewage is dumped into rivers, lakes and oceans, making beaches unfit for swimming and water too filthy to drink.

Pesticides don't decompose: they seep down into groundwater, making well water unsafe to drink, or get washed into lakes and rivers. They settle on the bottom for a while, then re-enter the system if the river bottom is dredged.

Garbage waste buried in landfill sites can pollute the groundwater and nearby rivers and lakes.

Factories discharge water contaminated with chemicals into rivers. Heated water dumped by factories also kills fish because they can't tolerate the higher temperature of the water.

Fertilizers on farmers' fields are washed into rivers, causing algae to multiply and choke off aquatic life. The rivers dump the fertilizers into lakes and oceans.

Wildlife alert

Water pollution is dangerous to every living creature — to humans who need clean drinking water and to plants and animals that live in or near water.

Suppose you stir a few drops of food colouring into a large container of water. The colouring quickly disappears, but the tiny molecules are still in the water even if you can no longer see them. Imagine that the food colouring were toxic (deadly) chemicals dumped into a river. As the river carries the water away, the molecules spread farther and farther apart, affecting the plants and animals that live in the river.

A little pollution can travel a long way. Even a container of water as huge as the ocean can become polluted — with deadly results. First the tiny plants of the sea, plankton, absorb a small amount of pollution. This ends up in the tiny animals that feed on plankton. These tiny creatures get eaten by larger fish, which are themselves eaten by even larger fish.

The pollution builds up through the food chain so that eventually even the biggest animals end up with large amounts of toxic chemicals in their bodies. The most polluted water animals are believed to be the beautiful beluga whales living in the St. Lawrence River. When they die, their bodies are so full of deadly chemicals that they are considered dangerous toxic waste and disposed of with great care.

Humans at the top of the food chain can also be victims of water pollution. Some people living around Minamata Bay in Japan in the 1950s went blind and deaf or died after eating fish poisoned by mercury.

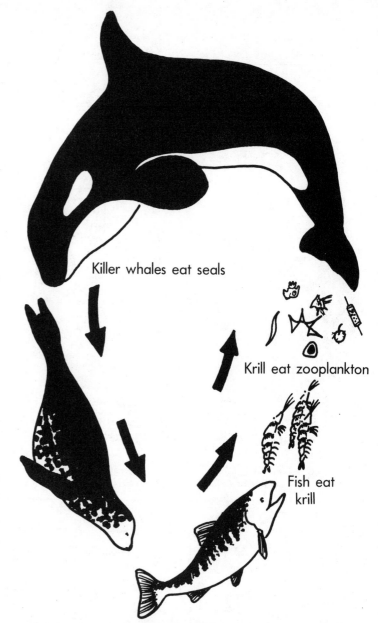

Killer whales eat seals

Krill eat zooplankton

Fish eat krill

The mercury and other wastes were dumped into the bay by a nearby factory. Native people in northern Ontario have also become victims of mercury poisoning brought about by dumped waste water. Other toxic chemicals have ruined water from wells; to avoid getting sick, people must buy bottled water.

Discover how pollution threatens wildlife by trying these experiments.

80

Danger 1: Phosphate power

Detergent containing phosphates might make great suds, but it can harm wildlife when it goes down the drain. Try this and see why. The experiment takes two to three weeks to do, but it's well worth the wait.

You'll need:
pond water
pond insects
pond plants, such as elodea
mud
5 jars the same size
a large bowl
a large spoon
dishwashing detergent containing
  phosphates
an eye dropper

1. Collect enough pond water, insects, plants and mud to fill each jar. Put the same number of plants and insects in all the jars and the same amount of mud from the bottom of the pond.

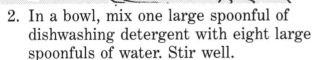

2. In a bowl, mix one large spoonful of dishwashing detergent with eight large spoonfuls of water. Stir well.

3. Using the eye dropper, add different amounts of detergent to each jar. Add one eye dropper full to the first jar, two to the second jar and so on. Add nothing to the fifth jar. Label each jar with the amount of detergent you put in.

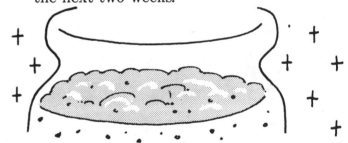

4. Put all the jars in a sunny windowsill and watch what happens to them over the next two weeks.

Did some of the plants and insects die? Did a brown scum appear on the surface of the water in some of the jars? Phosphates in the detergent overfed the algae, tiny one-celled plants that live in pond water. They formed a mat, or "bloom," on the surface and blocked out the sunlight for the plants underneath, so they died. Bacteria that have the job of breaking down dead plants and rotting algae took over and used up the oxygen in the water, so all the other organisms and animals died, too.

Making water clean to drink

Water in a lake or river may look crystal clear, but don't be fooled. Usually it's not fit for humans to drink. Different things can affect water quality, taste and smell. Brownish water, for example, contains particles of soil, sand and rocks that are washed into the water by rain or by rivers. Water that is green and murky contains tiny organisms that provide food for fish and plants. Some of these organisms are bacteria that decompose (break up) leaves and dead animals. This is nature's way of recycling things. Over many years, the water eventually cleans itself, as long as humans don't upset the process.

But when people begin dumping huge amounts of garbage and sewage (what goes down the sink and toilet) into the water, the water becomes too polluted for nature to clean. Soon the balance of nature is upset. Plants and fish start to die as other organisms take over and use up all the oxygen. The water smells foul, looks terrible and makes people very ill. Who wants to drink water with toilet refuse in it? To make this polluted water safe to drink, it must be cleaned and purified at a water treatment plant. See page 76 to see how water is purified.

82

Where's the bacteria?

Does your local water contain bacteria? Try this to find out.

You'll need:
4 small containers to collect water samples
paper and felt-tip markers
tape
4 clean glass jars, sterilized if possible
  (ask an adult to help with this)
a large potato
a saucepan
a teaspoon

1. Collect water in separate containers from four different places: from a pond, mud puddle, rain bucket and the tap. Tape on labels to show where each sample came from.

2. Put a few spoonfuls of pond water into a jar and label it. Do the same for the other three water samples.

3. Boil the potato in a saucepan on the stove for about 20 minutes. While it's still warm, cut it into four large slices. Put a slice into each glass jar.

4. Put the jars in a dark, warm place so the bacteria will grow on the food.

5. After a few days, check the jars to see which potato slice has the most bacteria growth on it. The slice with the thickest growth shows there were many bacteria in the water. If the potato slice in tap water has a lot of fuzzy growth, you should get your water tested. Tap water should *not* contain bacteria!

Lake Erie — the Great Lake that almost died

In the 1960s and 1970s, people became very worried about Lake Erie, the smallest and shallowest of the Great Lakes. The water smelled terrible and looked murky and soupy. Dead fish and rotting algae washed up on the beaches, which had to be closed to swimmers. The lake's surface was covered by large mats of blue-green algae. Tests showed that there was no oxygen at lower depths of the lake.

Lake Erie was dying because of a process called eutrophication. This happens when a pond or lake is overfed with phosphorus and nitrogen. Sewage dumped by cities, waste water dumped by factories and fertilizers from fields all contain phosphorus and nitrogen, as

well as deadly chemicals. Since 1972, laws have been passed in Canada and the United States to reduce the amount of these chemicals going into the water and to save the lake from dying.

Lake Erie has recovered, but the massive dumping of chemicals and other wastes in the Great Lakes and elsewhere poses a serious threat to the drinking water of millions of people.

Danger 2: Plastic kills

Take an elastic band and put it around your wrist. Can you get the elastic off easily without using your other hand or touching anything? Imagine the problem a fish has removing an elastic band or plastic ring from around its body!

Plastic bags or rings from six-pack cans are often mistaken for food by water animals. When they try to eat the plastic, they get entangled and die. Other animals (such as sea turtles) *do* eat plastic bags, thinking they're jellyfish. The World Society for the Protection of Animals estimates that plastic garbage kills more than two million birds, turtles, whales,

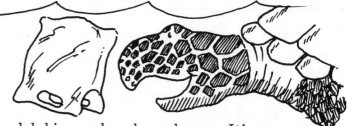

dolphins and seals each year. It's now illegal for ships to dump any plastic garbage overboard in the ocean.

What can you do to help the sea animals? Don't buy six-pack pop cans attached together with plastic rings. (If you do find plastic rings on the beach, cut them up into tiny pieces so that they won't end up strangling sea birds or animals.) Try to reuse shopping bags so that fewer get thrown out.

Danger 3: Hot water blues

If you leave a glass of cold water on a table for a few hours, you'll notice that tiny bubbles of air form inside the glass. The bubbles contain oxygen that is escaping as the liquid warms up. Why? Warm water can't hold as much oxygen as cold water.

Hydroelectric and nuclear power plants require huge amounts of cold water for cooling. Cold water from a nearby lake or river is pumped into the plant. As the cold water does its job of cooling the machines, it heats up. When the hot water is dumped back into the lakes, it causes the temperature of the surrounding water to rise. The result? Fish and plants that are used to cold water are killed off. There isn't enough oxygen in the warmer water to keep them alive.

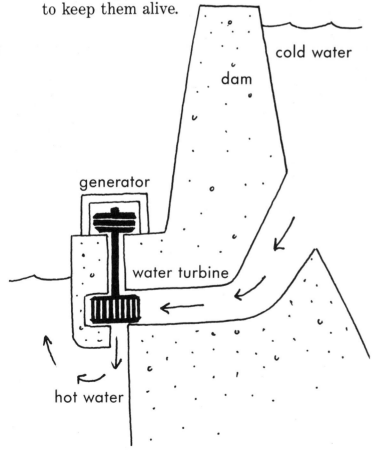

Danger 4: Oily waters

What happens when an oil tanker leaks or spills its cargo? For many water animals and birds, the answer is deadly. Try this and see why.

1. Pour a few drops of salad oil into a glass of water. Stir it well with a spoon.

2. Put the glass aside for 20 minutes. What happens to the oil?

Suppose that an oil film covered part of a lake or ocean. What would happen to seabirds? Many of them would die — they would be poisoned by the oil-tainted water or coated in oil.

Huge oil spills from leaking ocean tankers don't just kill birds. The oil acts like a blanket affecting wildlife for kilometres (miles) around. When the thick oily sludge washes up on the beach, it kills all the wildlife. And when the oil sinks down into the water, it kills the creatures below, even those on the bottom.

Danger 5: A vinegar bath

Do you like the sour taste of vinegar on salad? Why does vinegar taste sour? It's an acid. Imagine the vinegar-water solution in this experiment is acid rain. Would plants like it? Try it and see.

You'll need:
vinegar
a soup spoon
a measuring cup
2 small plants (the same kind)

1. Water one plant with tap water.
2. Water the other plant with vinegar water. To do this, put a spoonful of vinegar into 250 mL (1 cup) of water and use as much of this mixture as you need. If you run out of vinegar water, mix up some more.
3. Water both plants with their different mixtures for two weeks. What happens to the plant that gets only vinegar water? Do the leaves wilt? Eventually this plant will die because it can't tolerate the vinegar bath. See if you can save its life by repotting it in different soil and giving it clean water from now on.

Acid rain is a bit like the vinegar water. It can harm the plants on which it falls.

Acid rain is the name given to water in the atmosphere — including fog, sleet and snow — that gets polluted by dirty air. Exhaust from cars and factories that burn fossil fuels (such as coal) contains gases that become even more dangerous when they mix with rainwater. The rain falls on ponds, lakes and ocean coastlines and changes the chemical make-up of the water, so that fish and plants begin to die. Acid rain that falls on certain types of soil can kill the trees and plants growing there.

Are there ways to reduce acid rain? The solution lies in reducing air pollution. Factories that burn oil and coal can install devices called scrubbers to clean the smoke as it goes out the chimney, making the air less polluted for hundreds of kilometres (miles) around. Cars are bad news, too. Exhaust from car tailpipes is a major contributor to air pollution and to acid rain. People have to learn how to rely less on their cars if they want cleaner air and cleaner rain.

What you can do

Are there things you can do to save water? A good place to start is in your home. You can use less water by following the tips on pages 74 and 75. Here are some ways to reduce water pollution:

- Check the labels on detergents. Use ones that are phosphate-free, and use less detergent than the label says.

- Persuade your family to buy toilet and tissue paper that hasn't been bleached white with chlorine or dyed a fancy colour. The bleached paper decomposes but the chlorine and dyes stay in the water, affecting wildlife.

- The next time you're about to pour something down the drain, ask yourself if you'd want to *drink* the water mixed with chemicals from that product. This applies to art supplies, paint thinners, cleaning products and bug sprays.

- Do your part to clean up the air — use your bike or the bus rather than ask your parents to drive you everywhere.
- Buy drinks in returnable bottles or recyclable cans, not plastic containers that land in the garbage pail or in the river or beach. Collect the bottles and return them to the store. Never leave empty bottles at the beach or park.

- Declare war on garbage. Be cool — take a lunch pail to school, not a bag. The less garbage that goes to a landfill site, the less chance nearby water will get contaminated.

Here are some projects you can do at school:

- Suggest that your class adopt a local pond, stream or beach. Organize a litter clean-up day and invite parents, local politicians and newspaper reporters. Make a list of all the garbage the group collects and use this to write a report for your school newspaper. If the reporters don't come, take along a camera so that you can send pictures of all the litter to the local newspaper.

- Turn off water fountains. Report leaking faucets and blocked toilets to your teacher. Ask the principal if the school has replaced old lead water pipes, which leak harmful lead into the water all the kids drink.

- Does your town or city dump its sewage directly into the lake or ocean? Are the trees and water in your area affected by acid rain? These are questions you and your classmates could investigate for a science project or science fair.

- Write a letter about water pollution to your local politician, your city or state office of the environment, the governor of your state or the President of the United States. Be clear and polite, and don't forget to include your return address so that you will get a response. The more letters politicians receive about water quality, the more attention they'll have to pay to this problem. Ask your librarian to help you find out the addresses of your governor and other state officials. Here is the address of the President.

President of the United States
The White House
Washington, D.C. 20501

5. Wacky water games & tricks

Imagine playing soccer in a swimming pool with a melon as a ball. Sound crazy? It isn't to people looking for some water fun. Swimming-pool soccer players even grease a melon with vegetable oil to make it extra slippery. What kind of melon do they use? A <u>watermelon</u>, of course.

You've heard of ice hockey, field hockey and ball hockey, but what about underwater hockey? It's played in the deep end of a swimming pool by experienced adult divers. Why can only excellent divers play? They have to be able to dive for a brass puck, shoot it along the pool's bottom with a small curved stick and score!

Yes, people do some pretty wacky things in water. According to the *Guinness Book of World Records*, two certified SCUBA divers in 1981 rode a tricycle almost 104 km (65 miles) on the bottom of a pool in Tucson, Arizona. Why? To raise money for a charity. They stayed under for 60 hours. Then there was a German named Fritz Weber who walked on water for more than 297 km (185 miles) in 1983. He wore large shoes with floats.

You don't always have to get wet to have some water fun. Here are some easy tricks and games for you to try out by yourself or with friends.

Centre the cork

This trick is guaranteed to drive your friends crazy! All you need is an empty glass, some water and a cork.

1. Fill the empty glass almost to the top with water.
2. Ask your friends to float the cork in the centre of the glass. No matter how hard they try to centre the cork, it will drift to one side.

3. Gently pour in some more water so that the surface bulges over the top edge of the glass. Place the cork in the water. Presto! It moves to the centre where the water level is highest — on top of the bulge. The water forms a convex surface — the bulge — because of surface tension.

A water-drop lens

Have you ever tried using water drops to magnify small print and pictures?

You'll need:
sticky tape
a piece of paper with small type and
    pictures (a used stamp is ideal)
a small Styrofoam tray
plastic wrap
an eye dropper

1. Tape the piece of paper you want to magnify onto the Styrofoam tray. Cover the tray with plastic wrap so the paper won't get wet.
2. With the eye dropper, place a few drops of water on the plastic cover.

Experiment with different sizes of drops. Does the size of drops change how they magnify? Water drops curve up like the curve on a magnifying glass. They are miniature lenses.

Make a siphon

Challenge a friend to empty the water out of one bowl into another using only a piece of plastic tubing. This is a valuable trick to know how to do, especially if you ever need to empty water out of a fish tank or a swimming pool.

You'll need:
a bowl full of water
an empty bowl
plastic tubing (available at pet stores)

1. Place the bowl of water on top of a few books. This bowl has to be higher than the empty bowl. Place the empty bowl below.
2. Put your finger over one end of the tubing. Pour water into the other end.
3. Put one end of the tubing under the water's surface in the higher bowl. Put the other end in the empty bowl.
4. Release your finger and watch the water flow down the tubing.

Water words

The prefixes "aqua" and "hydro" both mean water. Try matching the words below to their meanings. Answers on page 103.

1. aquanaut
2. hydrophone
3. aqueduct
4. hydroponics
5. aqualung

a. man-made channel for transporting water
b. a portable diving apparatus strapped to a diver's back
c. an underwater explorer
d. an instrument for detecting sound by water
e. growing plants in water

inbows in the dark

You can see this rainbow only in the dark.

You'll need:
a small pocket mirror
a pie dish filled with water
a flashlight

1. Rest the mirror against the inner side of the dish as shown.

2. Turn the lights off. Hold the flashlight so that its beam hits the mirror. What do you see on the ceiling?
The water acts like a prism that bends, or refracts, the different wavelengths of light. Each wavelength has a different colour.

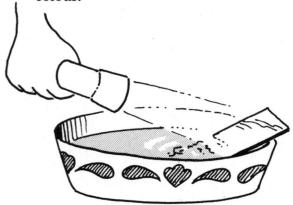

Water-balloon toss

Tossing water balloons is a fun way to cool off on a hot day. Fill empty balloons with a bit of water and then blow them up. (It's no good filling the balloons with too much water because they'll burst immediately.) Wear bathing suits.

Two players stand close together and begin throwing a water balloon back and forth. Each time you catch the balloon, take a step backwards. The loser catches the bursting balloon.

Floating needle

How can you make a needle float on water?

You'll need:
a drop of vegetable oil
a straight needle
a small piece of tissue paper
a bowl of water

1. Rub a little oil over the needle.

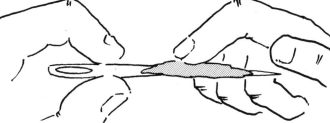

2. Place the paper carefully on top of the water.

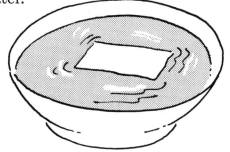

3. Gently rest the needle on the paper. What happens to the paper and the needle?

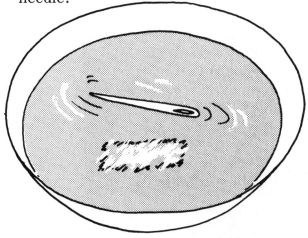

The paper sinks because it quickly absorbs water through its tiny tubes or fibres. Why does the needle float? Although it's made of steel and should sink, it's held up by the invisible skin on the water's surface. (For more about water's invisible skin, see page 25.)

Swimathon

- The youngest swimmers to cross the English Channel were a girl and a boy, both aged 12.
- Cindy Nicholas of Ontario was the first woman to cross the English Channel twice in 1977. She beat the men's record by 10 hours.

- The world's largest swimming pool is in Casablanca, Morocco. It's 480 m (1574 feet) long and is filled with salt water.

Make a bottle whirlpool

This is an impressive toy to show off to your friends. Even parents will want to play with it.

You'll need:
2 king-size plastic pop bottles
food colouring
brown tape (the kind movers use to seal
  boxes)

1. Fill one of the bottles half full of water. Add a few drops of food colouring to create a good effect.

2. Put the other bottle on top and tape where the two bottles meet.

3. Grasp the bottles at the taped joint and then rotate the toy around in a circle to make a waterspout. Watch how the water rises and falls in the two bottles.

What's happening? When you rotate the bottles, the water starts moving and forms a vortex. A vortex is a twisting spiral, which lets the air get in and the water flow down. Both things happen at the same time. P.S. You can watch water making a vortex when it goes down the drain.

Q. What's the fastest way to make water come out of a large plastic bottle?

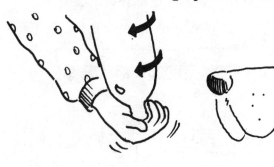

A. Plug one end of the bottle with your hand, then turn the bottle upside-down and rotate it hard. Remove your hand and watch the water pour out into the sink. The water will come out in a fast stream, rather than a slow glug-glug.

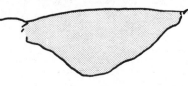

Magic glass

This is a terrific trick, but you'd better practise it over a sink or bathtub a few times before showing it off to your friends. The key is to make sure the water seals the cardboard to the rim of the glass.

You'll need:
a glass
a small piece of cardboard bigger than the glass's rim

1. Fill the glass with water to the very top. Make sure the rim is damp, too.

2. Put the cardboard on top of the rim.
3. Hold the cardboard in place with one hand and turn the glass upside-down with your other hand. Let go of the cardboard.

Why doesn't the water pour out? The pressure of air pushing up on the bottom of the glass is greater than the weight of the water pushing down.

Magic comb trick

You can make running tap water bend. All you need is a plastic comb and a sweater. The trick is to make use of static electricity, which your sweater will supply.

1. Rub the comb briskly on a sweater sleeve to give it an electrical charge.

2. Turn on the tap so that a small stream of water pours out.
3. Hold the comb close to the stream and watch the water bend towards the comb. Why does the water move? The electrical charge on the comb attracts the water.

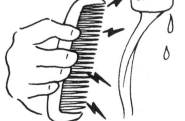

Water music

In 1761, the American scientist and politician Benjamin Franklin invented an unusual musical instrument called the armonica, or harmonica. A series of glass bowls were placed sideways on a spindle. As the spindle turned, the rims of the glasses touched water. Music was produced by rubbing a finger over the wet rims. Wolfgang Mozart composed music in 1791 for five instruments (a quintet) that

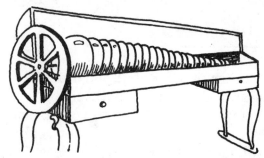

included the harmonica. Some modern composers have written music for glass tubes filled with different levels of water!

Would you like to start your own water orchestra? Begin by making one glass sing.

1. Fill a wine glass half full with water — the lighter and thinner the glass the better.

2. Wet your middle finger and hold it over the rim of the glass as shown. Now rub your finger around and around the glass's rim. Keep practising until you get a clear sound.

3. Once you've mastered one glass, line up eight identical wine glasses on a table. Keep a sponge and jug of water nearby.

4. Pour a small amount of water in the first glass, a larger amount in the second glass and so on. The eighth glass should be almost full.

5. Rub your wet finger around the rim of the full glass first to get the lowest note. Then adjust the level of water in the next fullest glass to make the second note on your scale. Repeat until you have a full scale of eight notes.

6. Pour a different drop of food colouring in each glass for a more spectacular effect.

How is the music made? The vibrating glass produces sound waves that travel through the air. As you add water, it slows down the vibrations, making lower sounds.

Make a bottle band

Repeat the previous experiment with eight empty pop bottles lined up in a row. Fill the first with about 2.5 cm (one inch) of water, the second with 5 cm (two inches) and so on. Now blow across the top of each bottle to produce a sound. Which bottle has the lowest note this time?

Why is the scale reversed? The sound here is produced by the air vibrating in the bottle, not by the glass vibrating. The bottle with the most water has the least amount of air that can vibrate. This bottle has the highest sound, because the sound waves are vibrating faster.

Marina, the Water Detective

Marina's my name and water's my game. I'm a marine biologist and a pollution expert who specializes in solving water problems. Usually people who want to hire me call my office at the harbour and leave a message on my phone machine. If I'm not on a case, I like taking my submersible out for a plunge into the depths of the lake to catch the best show of all — the fish. It sure beats Nintendo and TV.

Lately I've been really busy. Here's what I did last week. Can you figure out how I solved the problems? Turn to page 103 for the answers.

August 5: The half-empty swimming pool

The Charbonneau family hired me to investigate why their pool had lost one-third of its water during their two-week vacation. No one had tampered with the pump, located in a locked garage beside the pool. The pool hadn't sprung any leaks. Could all that water evaporate in just two weeks? Quite frankly, I was stumped. Then I noticed a hose leading out of the pool down a hill.

"Ah hah!" I said. "Your neighbours have played a trick on you." How had the neighbours emptied one-third of the pool?

August 7: The weed-filled pond

The Sewells asked me to drive out to their farm to look at their pond. They were wondering why the pond had become so overgrown with green algae and thick weeds. The kids didn't want to swim in the pond any more because of the weeds and algae. I had to agree — all that green stuf made swimming unappealing. Phewee!

Like the other farms in the area, the Sewells had a septic tank buried underground not far from the house. The septic tank stored all their waste water from the toilet and sinks. The pond was located beside a large corn field. Why was the water so yucky?

August 9: The smelly lake

I was surprised to get a call from the
Little Lake Cottagers' Association because
I'd worked with this group the year before
to clean up the water. Since then a new
marina for boats had been built on the lake.
The cottagers were hopping mad because
their beaches smelled like rotten eggs, and
oil was washing up on their sand. They'd
already checked all the septic tanks
surrounding the lakes, so what could be
dumping sewage into the water? And
where did the oil and litter come from?

August 10: Prisoner escapes from bathroom

My friend, Detective Singh, asked me to
help the town police figure out how a
prisoner had escaped from the local jail's
bathroom. The night before the guard had
let the prisoner use the bathroom and
forgotten to escort him back to his cell.
The next morning the guard had to break
through the bolted door and found the
bathroom empty. Look at the picture below
and figure out how the prisoner escaped.

On Friday I finally got to go down in my
submersible to about 200 feet. I was just
settling back to enjoy the scene when I
spotted four pop bottles, a tire, a car and
some tin cans on the lake bottom. All that
garbage spoiled my ride.

101

Glossary

acid rain: the name given to water in the atmosphere that is polluted by dirty air. The air gets polluted when invisible gases are released by cars that burn gasoline and power plants that burn coal. Some of these gases make the rainwater acidic. Acid rain or snow then falls down on ponds, lakes and plants and pollutes the water and land.

algae: plants that live in fresh or salt water.

aqua: the Latin word for water. *Aquatic* means to do with water. An *aquatic* animal lives in water.

aquifer: a huge underground layer of porous rock that holds water.

atmosphere: the blanket of gases surrounding the Earth.

cells: the basic units of living things. Cells are filled with a water-like liquid.

chlorine: a gas often added at a water-treatment plant or in a swimming pool to kill germs.

condensation: the process of changing from a gas to a liquid.

continental shelf: the undersea ledge of land around continents.

current: a body of water that flows through a lake or ocean.

dissolve: when a solid (such as salt) or a gas (such as oxygen) disappears in a liquid, it is said to have dissolved.

drought: a long period of time with little or no rain.

evaporation: the process of changing from a liquid to a gas.

fresh water: water that lies on the earth's surface in lakes and rivers or under the ground in aquifers.

groundwater: water that collects under the ground in porous rock. This is the water we pump up in wells and use to irrigate fields.

hydro: a word meaning to do with water. *Hydro*electricity means electricity generated from the energy of moving water.

molecule: a tiny piece of a material containing two or more atoms, the tiniest particles.

oxygen: a gas found in air and water that is necessary for plant and animal life.

pesticide: a chemical preparation that kills insects.

phosphates: a fertilizing substance containing phosphorus that makes plants grow; often found in detergents.

plankton: tiny plants and animals that float near the surface of oceans and lakes. Plankton are food for many fish, whales and sea animals. Plankton thrive in cold water, which is why the north Atlantic is greener than a tropical sea.

pollutant: a substance that pollutes, or dirties, the water or air, making them impure.

pollute: to make something too dirty and unhealthy to use. Polluted water containing raw sewage or dangerous chemicals is said to be contaminated, or impure.

sea: a smaller division of an ocean, usually partly enclosed by land (such as the Mediterranean Sea). Sometimes the ocean is also called the sea.

sewage: includes human wastes, chemicals and other things flushed down the toilet or thrown into rivers.

surface tension: the stretchy "skin" of a liquid caused by the attraction of the surface molecules to one another.

tides: the rise and fall of the oceans twice a day.

water cycle: the circulation of water from the oceans and lakes to the atmosphere and back to land and the ocean.

water table: the topmost surface of groundwater which wells tap for their water.

water vapour: the invisible gas formed when water is heated.

Answers

Thirsty plants, *page 11:* Three to four bathtubs full of water a day.

Caves and salt icicles, *page 23:* 1200 years, a metre (yard) every 40 years.

Science teasers, *page 33:*

1. The best time to water is in the early morning or at night, when less water will evaporate into the cool air. During the day, the heat of the sun speeds up evaporation.

2. Pour some salt on the ice cube to make the ice on the surface melt, then press the string down onto the ice with your finger for a few seconds. Take away your finger. The water will freeze again over the string, so you can lift up the cube.

3. It's easier to lift someone up in water because of water's buoyancy. Buoyancy is a force that pushes upwards in a liquid.

4. Roll the Play-Doh into a solid ball to make it sink. The ball shape will make it heavier, or denser, than water. Sculpt the Play-Doh into a boat and it will float. Why? Its new shape displaces (pushes aside) more water than the ball shape.

5. Fill a measuring cup with 125 mL (½ cup) of water. Then add spoonfuls of margarine until the water level measures exactly 250 mL (1 cup).

6. The water level will stay about the same after the ice cube melts because water from the ice takes up less space than the ice itself. Remember: when water freezes and changes to ice, the ice takes up about one-tenth more space than liquid water.

What is it?, *page 48:* 1 hail; 2 frost; 3 snow; 4 fog.

Which sea monster am I?, *pages 56-57;* 1. giant squid; 2. shark; 3. blue whale; 4. manta ray; 5. coelacanth; 6. giant octopus; 7. oarfish; 8. dugongs.

Rivers, *page 61:* President Gerald Ford grew up in Grand Rapids.

How river smart are you?, *pages 62-63:* 1 d; 2 h; 3 c; 4 f; 5 b; 6 i; 7 e; 8 j; 9 a; 10 k; 11 g.

An airplane view of water, *page 64:* 1 a; 2 c; 3 b.

Water words, *page 93:* 1 c; 2 d; 3 a; 4 e; 5 b.

Marina, the Water Detective, *pages 100-101:*

August 5: They used a hose to make a siphon. For more about siphons, see page 93.

August 7: The septic tank was leaking. Phosphates in the detergent from the sink were fertilizing the aquatic plants, making the algae grow rapidly. Fertilizers on the fields had run off into the pond.

August 9: The boats were dumping raw sewage, oil and garbage into the lake.

August 10: He turned on the tap, flooded the bathroom and escaped through the window at the top!

Index

Here are some more science-activity books like *Water Science:*

Balloon Science, by Etta Kaner

Bursting with over 50 experiments and a pack of colorful balloons

$8.95, 0-201-52378-7

Sound Science, by Etta Kaner

Resounding with more than 40 sonic experiments and activities

$8.95, 0-201-56758-X

The Amazing Books look at favorite foods and important substances, with scientific and historical facts and dozens of projects for home or school:

The Amazing Apple Book, by Paulette Bourgeois

$6.95, 0-201-52333-7

The Amazing Egg Book, by Margaret Griffin and Deborah Seed

$6.95, 0-201-52334-5

The Amazing Milk Book, by Catherine Ross and Susan Wallace

$6.95, 0-201-57087-4

The Amazing Potato Book, by Paulette Bourgeois

$6.95, 0-201-56761-X

The Amazing Paper Book, by Paulette Bourgeois

$6.95, 0-201-52377-9

The Amazing Dirt Book, by Paulette Bourgeois

$6.95, 0-201-55096-2

Addison-Wesley Publishing Company
Reading, Massachusetts 01867-3999